WHY A.I CAN'T STEAL MY JOB

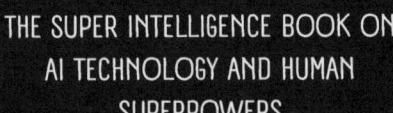

THE SUPER INTELLIGENCE BOOK ON
AI TECHNOLOGY AND HUMAN
SUPERPOWERS

OFEIMUN FAVOUR

WHY A.I CAN'T STEAL MY JOB

For inquiries, reviews, or special discounts for bulk purchases, please contact the author at: ofeimunfavour@gmail.com

TABLE OF CONTENT

Acknowledgement

Introduction

Chapter 1

The Evolution of Work and the Human Mind's
Path of Least Resistance

Chapter 2

Understanding AI and Robotics

Chapter 3

The Impact of AI and Robotics on Jobs

Chapter 4

The Magic of Human Ingenuity and Creativity

Chapter 5

The Irreplaceable Human Touch

Chapter 6

Education to Thrive in an Automated World

Chapter 7

The Career of the Future

Chapter 8

Recommendations and ethical considerations

Conclusion

Notes

Index

ACKNOWLEDGEMENT

Life bestows upon us many gifts, some evident and others hidden, waiting to be discovered. Among these, one stands out as the cornerstone of our existence, an anchor in the tumultuous seas of life — the gift of FAMILY. Each day, I am enveloped in gratitude for the familial bonds that have been my guiding light.

First and foremost, my heart swells with gratitude to GOD, the ultimate giver of life and purpose. His unwavering faithfulness has been my compass, guiding me through every challenge and triumph.

A special note of appreciation goes to LAZARUS UAMAI, my brother-in-law, whom I affectionately address as my uncle. His selfless dedication to my journey has been nothing short of monumental. His wife, my cherished sister EHINOR UAMAI, has been a beacon of support from the very inception of this vision. Her unwavering belief and encouragement have been instrumental in bringing this dream to fruition.

To my parents, DR GODWIN OFEIMUN and DORA OFEIMUN, pillars of strength and wisdom, I owe my deepest gratitude. Their enduring prayers and indomitable convictions have been my inheritance, guiding my path through life's intricate maze. My siblings, JENNIFER OBASUYI, ONOSEJERE OFEIMUN, and OMOGBAI ALFRED, have been my steadfast allies, their support and encouragement a constant source of strength.

The genesis of this book can be traced back to a profound conversation with my best friend and the book's editor, GODSWILL IRIABIJEH. His unwavering support and insights have been invaluable in shaping this narrative. OLUWAFIMI FADAHUNSI, a dear friend turned brother, introduced me to the wonders of generative AI and has been a bedrock of support throughout this endeavor.

A heartfelt thank you to our esteemed interviewees, RANIA HAFEZ and THEO SSEMAKULA, who generously shared their insights and expertise. Their contributions have been instrumental in enriching this work.
Lastly, to my pastor and mentor, VICTOR ENWEREUZOR, who has been both a guiding light and a cherished friend, I extend my deepest gratitude.
To all those who have touched my life, named and unnamed in these pages, know that you hold a special place in my heart. Your influence has been immeasurable, and for that, I am eternally grateful.

INTRODUCTION

It began with a simple question, one that I posed to many: "Do you think AI can take your job?" The reactions were immediate and varied. Many became defensive, touting their creativity and unique skills, yet skirting around a direct answer. As a fashion designer, my work is deeply rooted in creativity. Yet, I couldn't help but acknowledge the repetitive tasks in my daily routine that could easily be automated. This internal debate, this self-reflection, became a persistent echo in my mind: could AI truly replace me in my profession?

My epiphany came unexpectedly. As I delved into the capabilities of chatbots like ChatGPT, I realized that the need for certain professions, like copywriting, could diminish in the face of advanced AI. This led me to a profound thought: is creativity merely an amalgamation of external experiences?

One day, in the midst of a voice call with a friend, the question resurfaced with clarity: "Can AI steal my job?" Internally, I responded with a resounding "Yes." But then, a counter-thought emerged, forming the foundation of this book: "WHY AI CAN'T STEAL MY JOB." Even as my friend continued speaking, my mind was awash with ideas, concepts, and a newfound determination to explore this topic further.

This book is not just a culmination of research and insights but a testament to the very essence of my initial query. While AI has the potential to replace certain tasks, it also holds the power to augment human creativity. Throughout the creation of this book, various AI tools were employed for editing, research, proofreading, and critique. Yet, the core ideas, the soul of this book, are undeniably human.

In the pages that follow, we delve deep into the harmonious interplay between human creativity and the ever-evolving realm of AI. We explore the myriad ways our innate creative essence distinguishes us in a world increasingly influenced by algorithms and machine learning. The concept of the "Element" is introduced, highlighting the magical confluence of natural talent and personal passion.

As we journey through the digital age, we challenge the narrative of human obsolescence in the face of AI. Instead, we champion the belief that AI, in its essence, is a tool designed to augment, not replace, human potential. Join me on this enlightening exploration as we chart a course through the intricate dance of humans and AI. Together, we'll uncover the pivotal role of human creativity in crafting a future where technology amplifies our potential, setting the stage for a world of boundless possibilities.

CHAPTER ONE

THE EVOLUTION OF WORK

The history of work and its evolution is a complex narrative, shaped by various factors including technological advancements, economic shifts, and societal changes. The book "Rise of the Robots" by Martin Ford provides a comprehensive analysis of the evolution of work. In the 17th and early 18th centuries, people primarily worked on farms, cultivating crops and raising livestock. This work was physically demanding, requiring significant time and effort. Most tasks were performed manually or with the aid of simple tools. In addition to agriculture, there were artisans and craftsmen who specialized in trades such as blacksmithing, carpentry, and weaving.

These workers often maintained their own workshops and trained apprentices to carry on their skills. Ford stated, the first Industrial Revolution, which began in the mid-1700s in Great Britain before spreading to other parts of Europe and the United States, brought about significant changes. This period saw remarkable advancements in manufacturing and production technologies, leading to a shift from an agrarian and handicraft economy to one dominated by industry and machine manufacturing.

The advent of new technologies and machinery made it possible to automate many manual tasks, leading to increased productivity. This also spurred the growth of factories and urbanization as people moved from rural areas to cities in search of work. The nature of work also underwent significant changes. Instead of being directly involved in the production of goods, many workers became part of a larger system, often performing repetitive tasks. This shift had significant social and economic impacts, leading to changes in labor laws and the formation of labor unions to protect workers' rights.

Meanwhile, the book "The Second Machine Age" by Brynjolfsson and McAfee provides some insights into how factories where operated and how people felt during the early stages of industrialization. factories were powered by steam engines, with machines clustered around the main power source.

The transition to electricity did not immediately change this layout, as factories simply replaced the central steam engine with large electric motors. It took about thirty years for a new generation of managers to redesign factories around the natural workflow of materials, with each machine having its own small electric motor. This change significantly increased productivity and set the stage for further improvements in manufacturing productivity through complementary innovations like lean manufacturing and steel minim ills.

There was also skill-biased technical change, where technology increases the demand for skilled labor while decreasing the demand for less skilled labor. Technologies like payroll processing software and automated inventory control have automated routine work, while technologies like big data and analytics have increased the value of people with engineering, creative, or design skills.

This shift led to a decrease in demand for less skilled labor and an increase in demand for skilled labor. However, this period of rapid change was not without its challenges and controversies. For instance, the book mentions the Luddite movement, where English textile workers whose jobs were threatened by automated looms rallied against mills and machinery. This suggests that there was significant anxiety and resistance among some segments of the population, particularly those whose livelihoods were directly threatened by the new technologies.

Brynjolfsson and McAfee also highlighted on the concept of a "second Industrial Revolution" that occurred between the late 19th and early 20th centuries, characterized by the introduction of electricity, the internal combustion engine, and indoor plumbing with running water. This period also saw significant growth in productivity, but growth slowed markedly after 1970, this slowdown was attributed to the exhaustion of the benefits of the great inventions of the second industrial revolution. Once the internal combustion engine ran out of fuel, there wasn't much left to drive growth at the same pace leading to debates about the future of innovation and economic growth.

During this era, the phrase "hard work" was held in high esteem. Success and growth were perceived as the fruits of laborious effort. If one was not toiling strenuously, they were labeled as lazy, and their chances of success were deemed slim. An online definition of hard work I encountered describes it as 'spending long hours on a task or series of tasks'. The Oxford Dictionary defines hard work as 'putting a lot of effort into a job and doing it well'.

These definitions suggest that the quality of hard work is determined by the amount of time and labor invested. In the modern era, we have introduced the concept of "smart work," which has gained widespread acceptance, particularly among younger generations. The same article that provided the definition of hard work defines smart work as 'completing the same amount of work (with the same quality) in less time'. This shift in perspective has given rise to phrases like 'don't work hard, work smart' and has fueled the increasing demand for cognitive skills. However, some schools of thought maintain that working smart doesn't necessarily preclude working hard. It's just that people have defined hard work by the limitations of their perceptions. Hence, hard work generally refers to the expenditure of energy (both mental and physical) to accomplish a task or goal, encompassing a wide range of activities, including manual labor, intellectual endeavors, artistic creation, or even athletic training.

By exploring the notions of hard work and smart work, we begin to delve into the human mind's path of least resistance when it comes to productivity and time management. The evolution of work and the changing perceptions of labor have influenced how individuals approach their tasks and seek optimization in their efforts. Understanding these concepts and the underlying motivations can shed light on why individuals are drawn to the efficiency and potential time-saving benefits offered by artificial intelligence and robotics.

In the following chapters, we will further explore the implications of these changes in work dynamics and delve into the specific sectors that are most likely to be affected by AI and robotics. We will examine the challenges and opportunities they present and propose strategies to navigate this evolving landscape. By understanding the human mind's path of least resistance and the shifting paradigms of work, we can begin to explore how AI cannot simply replace human jobs but instead augment and transform the way we work and contribute to society.

The Concept of Time: The Human Mind's Path of Least Resistance

Time is the essence of our existence, a currency we exchange for actions that fulfill our desires. We strive to save time, often at the expense of our own or others'. Consider a photograph: when we capture a moment, we are essentially freezing a slice of time, underscoring its profound significance. Time is the greatest healer, the silent force behind growth and change. In physics, time is a fundamental measure, underpinning concepts such as power, acceleration, and velocity. Humanity's natural inclination towards the path of least resistance is a testament to our subconscious valuation of time.

In his book, 'A Brief History of the Philosophy of Time', Adrian Bardon delves into the Enlightenment era of the 17th and 18th centuries. This period marked a shift in philosophical exploration from the nature of time to our understanding of it. Philosophers began to pose epistemological questions, delving into the origins and limits of our knowledge. Figures like John Locke were deeply interested in unraveling how our minds grapple with the concept of time. They sought to uncover the roots of our ideas about time, leading them to explore the human mind's perception and comprehension of time.

In more recent times, the study of the philosophy of time has evolved to encompass more complex questions. Scholars are now examining not just time in isolation, but also its relationship with space., giving rise to the concept of "space-time." Intriguing ideas such as time travel and the connection between time and the laws of physics are also being explored. This exploration underscores the profound significance humans attribute to time. We have always strived to optimize our existence, extracting the most value from the fleeting moments that make up our lives. This endeavor has given rise to various strategies and philosophies.

One such philosophy is eloquently presented in Greg McKeown's book, "Essentialism: The Disciplined Pursuit of Less." McKeown argues that by concentrating on fewer tasks and executing them well, we can manage our time more effectively, communicate more efficiently, and enhance the speed and quality of our decision-making. This philosophy aligns with the theory that humans are innately inclined to conserve time by pursuing the path of least resistance for productivity.

Our brains employ temporal processing as a means to conserve time. When presented with a stimulus, our brains generate multiple, inconsistent interpretations. One interpretation prevails because it offers the simplest or most useful explanation in most situations. This cognitive shortcut can be viewed as a method of "saving time" in terms of cognitive processing, enabling us to respond swiftly and efficiently to our surroundings.

The rise of technology has played a significant role in saving time. With digitalization and automation, we can now set our priorities right. From our homes to businesses, gadgets and equipment have been replaced to save time and increase productivity. The embrace of technology is evident, as we are quick to snap our fingers to adopt new products that promise time-saving and relief. This attitude is widespread among individuals and establishments alike. Every company strives to use minimal time to perform quality tasks, reflecting the shared desire to optimize efficiency and productivity.

Overall, the concept of time and our inherent inclination to save it through various means, such as philosophical exploration, adopting productivity strategies, and embracing technological advancements, demonstrates our deep-rooted appreciation for the value of time and our pursuit of the path of least resistance.

Looking Ahead: the rise of personal computers

Computers were developed to automate complex calculations and data processing tasks that were difficult and time-consuming for humans to perform manually. They were designed to increase efficiency, accuracy, and productivity in various fields, including science, engineering, business, and government. The late 1970s and 1980s marked the beginning of the personal computing era.

In 1975, the MITS Altair 8800 was released, often considered the first personal computer. However, it was the launch of the Apple II in 1977 and the IBM PC in 1981 that truly brought computers into homes and offices around the world. These machines were smaller, more affordable, and more user-friendly than their predecessors, opening up computing to a much wider audience. During this period, the development of software also accelerated.

The first widely used operating system, MS-DOS, was released by Microsoft in 1981. Graphical user interfaces, which allow users to interact with electronic devices through graphical icons and visual indicators, were introduced. Apple's Macintosh, launched in 1984, was one of the first personal computers to offer a graphical user interface.

Meanwhile, the 1990s brought about the age of the internet. While the internet had been around in some form since the 1960s, it was the invention of the World Wide Web in 1989 that made it accessible to the general public.

The web, along with the development of web browsers like Netscape Navigator and Microsoft's Internet Explorer, transformed computers from standalone devices to interconnected, global communication tools. This period also saw the rise of portable computing. Laptops, which had been around in some form since the 1980s, became more powerful and affordable.

The 2000s saw the introduction of smartphones and tablets, further changing the way we use computers. Today, we're on the cusp of another revolution in computing with the rise of artificial intelligence (AI) and quantum computing. AI is enabling computers to learn from experience and perform tasks that, until recently, required human intelligence. This includes everything from recognizing speech and images to diagnosing diseases and driving cars.

Quantum computing, still in its early stages, promises to bring about another massive shift in computing power. Unlike classical computers, which use bits as their most basic unit of information, quantum computers use quantum bits, or qubits, which can represent multiple states at once. This allows them to perform complex calculations much more quickly than classical computers.

The history of computers is a testament to human ingenuity and our relentless pursuit of knowledge and efficiency. As we continue to innovate and push the boundaries of what's possible, who knows what the future of computing will hold?

CHAPTER TWO

UNDERSTANDING A.I AND ROBOTICS

The Significance of Data in AI and Robotics

To comprehend AI and robots fully, it is crucial to first understand the concept of data, as it forms the foundation of these technologies. Data can be likened to the soul of an AI or robot; without it, they would be rendered useless or mere novelties. As a fashion entrepreneur, I have learned firsthand the importance of maintaining detailed records of cash flow, inventories, and customer information. These records serve as valuable tools for assessing strengths, weaknesses, and making informed decisions through forecasting.

For instance, keeping track of the number and types of shirts purchased in a month guides me in determining the quantity and variety of clothes to produce in the subsequent month. Analyzing the purchase patterns of customers online enables us to predict their preferences, thanks to analytic tools such as Google Analytics, which we integrate with our website. Through these records, we can gain insights into customers who browse our site and proceed to the checkout phase, even if they do not complete their purchase. Such information allows us to make informed decisions regarding marketing strategies and personalized advertisements. It is not uncommon for individuals to experience the phenomenon of seeing ads related to a topic they recently discussed.

This occurrence can be attributed to data records. For instance, my sister once believed she was being relentlessly pursued by an online store's ads after not purchasing a product she had browsed. I explained to her that the ads were not personal vendettas but rather marketing efforts based on her interest in the product. This situation arises from the permissions granted to certain apps to collect voice data, not as a result of magic, but as a consequence of recorded information.

The records we maintain, referred to as data, are a collection of facts, information, or statistics gathered for reference, analysis, or calculation. Data can take various forms, including numbers, words, measurements, observations, or even pictures. When processed, organized, structured, or presented in a specific context to make it useful, data becomes information. Humans have embraced the practice of gathering data across numerous realms, including behavior, weather patterns, medical records, communication systems, travel data, and work data. Similarly, AI and robots leverage data to make predictions, perform calculations, and make informed decisions, but at a faster and more accurate rate than humans.

Reflecting on my own journey as a fashion entrepreneur, I recall how managing customer demands and manually calculating daily cash flow were initially manageable tasks. However, as the company expanded, the line between meeting customer needs and maintaining records became increasingly blurred. Recognizing the need for assistance, I employed a secretary. As the business continued to grow, we transitioned from paper records to Microsoft Excel.

Today, specialized software exists to streamline record-keeping and facilitate data analysis, enabling us to make data-driven decisions. This exemplifies the power of data and how AI and robots harness it.

In the modern age, data has inadvertently become an abundant byproduct of our society, furnishing machines with an extensive library from which they can learn. Often, without realizing it, we generate and provide data to individuals, companies, society, and the internet. Daily actions such as clocking in at work, tapping a bus pass onto a train or bus, booking appointments with doctors or hospitals, and even the time spent on calls and the internet—all these actions are somehow stored by someone, a company, or the government. While the misuse of data can have negative consequences, when employed responsibly, it can greatly enhance life.

Consider the example of Apple. With approximately 80% of iPhone users in the US owning an Apple Watch, and a staggering 124.7 million iPhone users in the US in 2022, Apple has access to a wealth of data about individuals' health status through their Apple Watches. Leveraging this data, Apple has collaborated with health experts to develop health features that provide users with actionable insights into their well-being. Moreover, these innovations assist the research and medical communities in studying, tracking, and treating a wide range of conditions. The areas in which the Apple Watch is employed for study and tracking encompass heart rate and rhythm, cardio fitness, irregular rhythm notifications, ECG recording, AFib history, activity data, sleep patterns, blood oxygen levels, and more. By collecting and analyzing this data, Apple empowers individuals to gain valuable insights into their health while supporting advancements in medical research.

This example exemplifies the symbiotic relationship between humans and data-driven technology. We willingly provide computers with data pertaining to the areas in which they seek improvement, enabling them to learn quickly and contribute to our betterment.

AI and robots continuously refine their capabilities by processing data through techniques like machine learning. This approach allows machines to acquire information and skills by scrutinizing data, enabling them to become better at tasks through the relationships and patterns they discover. In essence, data is vital for AI and robots to evolve into more intelligent entities that excel at automating a wide array of tasks.

As we delve deeper into the realms of AI and robotics, we will uncover the various ways these technologies utilize data to enhance their functionality. From machine learning algorithms to sophisticated data analysis, AI and robots demonstrate an aptitude for processing vast amounts of data with speed, precision, and efficiency. Understanding the intricate relationship between data and AI/robots is essential for comprehending the transformative potential of these technologies in the workplace and society as a whole.

In the subsequent sections of this chapter, we will explore real-world examples of how AI and robotics are currently employed across different sectors. By examining these applications, we can gain a comprehensive understanding of the impact and implications of these technologies in various industries. From manufacturing and healthcare to transportation and customer service, we will witness the transformative power of AI and robotics in action.

Basic explanation of AI and robotics and their uses in various fields

Our lives and the workplace are changing in unimaginable ways thanks to artificial intelligence (AI) and robots, two representative technologies of our modern age. Although these two areas have blended into our vocabulary, they are sometimes interpreted incorrectly or seen in isolation. It is necessary to comprehend them holistically, understanding their connection and the specifics that characterize each field, in order to fully appreciate their potential and realize their implications.

AI, a branch of computer science, emulates tasks typically associated with human intelligence. It engages a wide array of activities from natural language processing and image recognition to decision-making and problem-solving. As demonstrated by IBM's Deep Blue, a chess-playing computer that defeated Garry Kasparov as the world champion in 1997, and by the widespread adoption of digital personal assistants like Siri and Alexa, which have become an essential part of many people's everyday lives, its real-world manifestations are pervasive and widespread. Simultaneously, AI has given birth to machine learning, a subset that trains algorithms on vast datasets, enabling predictions and actions. A prime example is Google's search engine, which uses machine learning algorithms to examine numerous web pages and deliver the most appropriate results to visitors.

However, designing, building, and using robots are the focus of the engineering discipline known as robotics. The enormous potential of this subject is shown in the wide spectrum of these automated machines, which vary from tiny drones and personal helpers to enormous

industrial apparatus and autonomous vehicles. Whether it's pipeline inspections, hazardous waste cleanup, or exact surgical operations, the primary benefit of robots is their capacity to perform activities that are too difficult, boring, or dangerous for humans to perform. Robotics and artificial intelligence frequently collaborate despite having different development paths. In reality, a lot of robots nowadays are powered by AI algorithms, creating a symbiotic relationship that increases the possibilities for advancement.
As AI-enhanced robots continue to push the envelope, a few fields that will profit from this fusion include autonomous transportation, environmental monitoring, and healthcare.

Numerous sectors are being revolutionized by this fusion of AI and robots. Adopting AI technologies can significantly increase efficiency by accurately and swiftly analyzing massive amounts of data, promoting informed decision-making, and automating repetitive operations. Parallel to this, robotics can speed up production, improve quality standards, and enhance workplace safety measures, demonstrating their disruptive potential in industries ranging from manufacturing to healthcare.

Although worries about losing jobs and managing clever machines still exist, optimism exceeds skepticism. Experts are optimistic that AI and robotics will create new job opportunities and foster human creativity and strategic thinking, suggesting that this technology transformation will spur societal growth rather than signal the end of a certain way of doing things. The spectrum of tasks that can be automated keeps expanding. Increasingly sophisticated AI systems are finding their way into complex domains, from predicting stock market trends with greater accuracy to providing personalized learning experiences in education.

In healthcare, AI is making it possible to identify patterns in symptoms and diagnose diseases much earlier, potentially saving countless lives.
When it comes to the creative arts, AI is even showing potential. Algorithms can now create music, generate unique artwork, and even write poetry, challenging our notions of creativity and artistic inspiration. However, it is important to acknowledge that while AI can mimic these tasks, the intrinsic emotional understanding and personal expression that humans bring to art remain unmatched.

In the domain of robotics, we are witnessing an equally intriguing evolution. Robots are becoming increasingly present in our daily lives, be it in the form of autonomous vacuum cleaners that keep our homes clean or drones that deliver packages right to our doorstep. Beyond that, robots are venturing into uncharted territories. For instance, in the field of space exploration, robots are being sent to other planets and moons to gather data and pave the way for potential human habitation. Robots are also proving to be a boon in addressing pressing environmental issues. Autonomous machines are being deployed to monitor wildlife, clean up oil spills, and even remove trash from the world's oceans. This not only highlights the versatility of robotics but also its potential to contribute to preserving our planet.

Furthermore, the fusion of AI and robotics holds great promise for the realm of personal assistance and caregiving. AI-powered robotic assistants are being developed to assist the elderly and differently-abled individuals, helping them with daily chores, providing companionship, and even monitoring their health.

This hints at a future where robotics and AI can contribute significantly to enhancing quality of life and fostering inclusivity.
While the convergence of AI and robotics offers countless possibilities, it also underscores the need for ethical considerations and regulations. Questions regarding privacy, security, and responsibility in the context of AI and robotics are increasingly critical.

As such, lawmakers, ethicists, technologists, and society at large need to collaborate and create a framework that ensures these technologies are developed and used responsibly. In the grand scheme, AI and robotics represent more than just technological advancements. They reflect our collective aspirations for a future where machines can complement human abilities, create new opportunities, and help address some of our most pressing challenges. As we venture deeper into this exciting frontier, it's vital that we do so with thoughtfulness and a firm commitment to harnessing these technologies for the greater good.

IMPACT OF A.I & ROBOTICS ON JOBS

The Changing Landscape of Job Sectors

The rapid rise of automation and technological advancements is reshaping job sectors in unprecedented ways. Throughout history, humans have adapted to evolving labor demands. We transitioned from agrarian work during the agricultural era to production jobs during the Industrial Revolution. As automation gained prominence, service-oriented roles emerged. Now, in the Information Age, we are witnessing yet another profound shift.

Let's take a moment to envision the traditional office space. Picture a desk adorned with a calculator, calendar, analog alarm clock, stacks of books, and a trusty typewriter. However, in today's modern office, many of these physical tools have been replaced by a single device—the computer. With software that serves as a calculator, alarm, typewriter, note-taking tool, and digital record-keeper, the computer has become a versatile and efficient companion. While certain components such as time clocks, calculators, alarms, and note-taking tools still have relevance, the integration of these functionalities into a single device has brought about convenience, productivity, and reduced stress. In a similar vein, jobs across various sectors are being transformed as automation enables the handling of multiple tasks and the provision of greater convenience.

Oxford University researchers conducted a comprehensive study on the future of work, revealing that nearly half of all jobs are at high risk of being automated by machines. One such sector experiencing significant changes is the financial industry.

Consider the introduction of Automated Teller Machines (ATMs) in the early 2000s. This innovation revolutionized banking by automating cash-handling processes and reducing the reliance on human tellers. Today, machine learning technology, powered by artificial intelligence (AI), is once again transforming the financial industry.

Natural Language Processing (NLP) has enabled computer systems to comprehend, interpret, and generate human language—a capability that proves invaluable for automated customer service and other tasks. The implementation of AI-powered tools and chatbots enables faster operations, cost-efficiency, and the provision of round-the-clock customer support. The financial industry is just one example among many sectors experiencing the impact of automation. Many factories throughout the world are deploying AI and robotic technology to replace humans in repetitive, unskilled work in the manufacturing industry.

For instance, 60,000 robots were placed by Foxconn, a significant Apple supplier, in their plants to perform welding and painting operations that previously needed a large number of human workers. It is relatively simple to automate a particular process, such as soldering a wire onto a circuit board or screwing two pieces together, but the task must be consistent over time and take place in a "regular" environment. For instance, the circuit board must consistently appear in the exact same orientation.

Businesses purchase specialized equipment for jobs like these, have their engineers programme and test it, then integrate it into their production lines. Production is forced to stop until the machinery is reprogrammed each time the task is altered—for instance, each time the location of the screw holes' changes. Additionally, many car manufacturers have implemented robotic systems to assemble cars, with Tesla using over 1,000 robots in its factory.

One practical example of AIs uptake in Manufacturing is that of ABB. ABB has been recognized as one of the pioneers in the field of AI. They have launched the ABB Ability solution that includes integrated field data to optimize the operation and maintenance of various industrial equipment and systems, reducing unplanned downtime and reducing maintenance costs.

This platform suggests corrective actions and reduces the risk of accidents. The system uses sensors to collect field data, which is then integrated and analyzed to give insights and solutions, allowing equipment operators to reduce unplanned downtime and maintenance costs. In addition to ABB, other manufacturing giants like General Electric, Kuka, and Siemens have long embraced the use of AI and are using it to automate various manufacturing tasks.

For instance, General Electric uses AI to advance its manufacturing process in its wind turbine plant. Kuka's robots are often used in car manufacturing, where they can do everything from welding to painting. Kuka has also been exploring the use of AI to make its robots more adaptable and capable of learning new tasks and Siemens uses AI to analyze data from sensors on its machines to predict when a machine will need maintenance. This allows Siemens to fix problems before they cause a machine to break down.

Automation in the Transportation Industry

One of the most visible applications of automation can be seen in the transportation industry, particularly with the development of self-driving or autonomous vehicles. These vehicles utilize a combination of sensors, algorithms, and machine learning to navigate and operate without human intervention. The potential of self-driving cars to revolutionize transportation is significant, offering reduced costs, improved efficiency, and enhanced safety by minimizing human errors.

I recall a fascinating experience when I ordered an Uber and was pleasantly surprised to find a Tesla car arriving to pick me up. As I entered the vehicle, the driver warmly greeted me, and I couldn't help but notice the unique door handle design. The driver kindly guided me on how to operate it, explaining that I should push the wider part of the handle with my thumb, causing it to pivot towards me. Once inside, I was greeted by a wide display screen that provided a 3D-like view of the car's surroundings, giving me a glimpse into the advanced technology at play. While the driver was still in full control of the car at the time, it became evident that in the not-too-distant future, it would no longer be surprising to see AI-driven vehicles comprising a significant portion of the vehicles on London's streets.

China, too, is experiencing a transformative shift in the transportation sector with the introduction of self-driving cars. Cities like Chengdu have already deployed autonomous taxis equipped with level 4 autonomous driving capability. These vehicles employ advanced technology, including multiple cameras and laser radar, to sense road conditions and create a real-time traffic status digital twin platform. This enables the AI system within the taxi to make informed decisions, improving both safety and efficiency.

The widespread adoption of autonomous vehicles has the potential to reduce costs by enhancing technology and minimizing labor expenses. Companies estimate that autonomous driving could lead to a halving of taxi fares, making transportation more accessible and affordable for a larger segment of the population. Furthermore, the elimination of the need for driver configuration in car design allows for mass production of sensors and computing units, ultimately driving down the cost of autonomous cars as the supply chain matures. With human errors accounting for 94% of road accidents, the implementation of automated cars aims to significantly enhance road safety.

Automation in the Aviation Field

The aviation industry has embraced the integration of artificial intelligence (AI) and data science to optimize various aspects of their operations and enhance efficiency. One significant application of AI in aviation is demand forecasting and revenue management. Airlines utilize predictive models that consider thousands of factors to accurately predict demand for different routes. By analyzing traveler behavioral data, such as abandoned searches on online travel agent sites or social media chatter, airlines can identify market gaps, raise fares for specific routes and dates to benefit from rising demand, and optimize revenue. They also employ ranking algorithms to match historical flight bookings with event data, enabling them to anticipate and respond to short-term spikes in demand driven by events like festivals or conferences.

Fuel optimization is another area where AI is making an impact. Airlines use predictive models to analyze fuel consumption data and consider factors such as fuel prices, the number of trips, and time periods.

This allows them to estimate the amount of fuel needed for each flight accurately, resulting in significant cost savings.

Airlines have implemented facial recognition technology to speed up boarding procedures and increase security. Airlines can optimize boarding times, speed up the boarding process, and cut down on waiting periods by scanning passengers' faces and comparing them to passport photos. By using neural network-powered picture identification techniques, the system also makes it possible to monitor plane preparation in real time. Airlines may ensure on-time departures and cost savings by taking preventative action by detecting issues that could cause delays. In order to speed up their boarding procedure, several airlines already provide biometric technology as a boarding option. One of these airlines is Delta Airlines, which has installed facial recognition software in numerous airports. In comparison to conventional boarding procedures, Delta has indicated that its passengers had a preference for biometric boarding of up to 72%.

Data science has proven instrumental in reducing food waste and driving profitability for airlines. By leveraging data analysis techniques, airlines can gain valuable insights into the demand for food items served on planes, allowing them to optimize their meal offerings and minimize waste. These models enable airlines to accurately predict the demand for specific food items during flights, thereby ensuring that they provide meals that align with passengers' preferences and reduce the likelihood of leftovers. A notable example is EasyJet, whose CEO, John Lundgren, recognized the potential of data science in addressing food waste and saving costs. Lundgren instructed EasyJet's data science team to analyze the demand for food items served on their planes.

By implementing the insights derived from the predictive models created by their data scientists, EasyJet has achieved substantial savings, amounting to millions of pounds annually, while concurrently reducing waste. EasyJet's data science team developed an algorithm that takes into account factors such as the time of flight departure and destination cities for different routes. This algorithm effectively predicts the demand for food items, enabling EasyJet to tailor their meal offerings to match the expected demand. By providing meals that are in high demand, EasyJet significantly reduces food waste and increases their profitability.

In the realm of air traffic management, AI is revolutionizing decision-making and conflict resolution for air traffic controllers. Projects funded by organizations like SESAR JU are developing AI-powered tools that analyze real-time data, predict potential issues, and provide optimal solutions. Transparency and trustworthiness are crucial considerations in integrating AI into aviation. Solutions are being developed to offer multiple levels of transparency, utilizing advanced visualization techniques such as heat maps. These tools support air traffic controllers in making informed decisions, providing a clear visual representation of complex data. AI algorithms can even estimate the likelihood of a pilot's go-around request based on historical data, allowing proactive measures to be taken to enhance safety and efficiency.

Artificial Intelligence in Health Care

Artificial Intelligence (AI) is revolutionizing the healthcare industry, bringing about significant changes and advancements. One healthcare professional, Theo, shared her experience as a nurse for over 15 years and highlighted the shift from paper-based documentation to digital documentation.

This transition has made accessing patient data and sharing information between healthcare professionals much easier and more efficient. Additionally, technology has played a crucial role in organizing patient records, leading to improved patient care. During the conversation with Theo, the topic of robots in healthcare arose. She mentioned that robots are being used in surgeries, controlled by surgeons. These robots offer greater flexibility than human hands, allowing them to access hard-to-reach areas during operations. However, it is important to note that robots still require programming by humans, and healthcare professionals need to undergo training to operate them effectively.

AI has shown significant potential in various medical specialties. In radiology, for instance, a deep-learning algorithm trained on over 50,000 normal chest images and nearly 7,000 scans with active tuberculosis has outperformed human radiologists. This AI system has the potential to expedite diagnosis, especially in regions where access to skilled radiologists is limited.

Dermatology has also benefited from AI, in its ability to identify skin cancer. A deep learning neural network trained with over 100,000 images correctly detected 95% of melanomas, surpassing dermatologists who accurately detected slightly over 86%. This advancement could lead to earlier detection of skin cancer, improving patient outcomes.

In oncology, IBM's Watson has generated interest for its ability to analyze vast amounts of data and provide treatment recommendations. While there are still challenges in fully integrating AI into complex medical decision-making processes, the potential for AI to assist in oncology is promising.

Cardiology has witnessed the potential of AI in predicting cardiovascular diseases. Deep-learning methods can analyze eye images to determine risk factors such as age, gender, smoking status, and blood pressure. Although further validation is needed, these studies highlight AI's potential in preventive medicine.

The COVID-19 pandemic brought about a significant revolution in the healthcare sector, particularly in the realm of telemedicine and remote care. Dr. Bertalan Mesko, a renowned Medical Futurist, emphasizes the transformative impact of the pandemic on healthcare in his video titled "How COVID-19 Changed The Future of Healthcare." According to Dr. Mesko, the pandemic has acted as a catalyst, driving the healthcare sector to adopt innovative digital health solutions and accelerating the evolution of remote care. Remote care has become the new normal, as the pandemic forced health networks to quickly embrace telemedicine. While regulatory concerns, patient privacy, and questions about compensating doctors have previously hindered the widespread adoption of remote care, the pandemic compelled health systems worldwide to rapidly implement telemedicine. Hospitals and governments across the globe embraced digital health, resulting in a significant increase in teleconsultations.

In Spain's Catalan region, teleconsultations skyrocketed from 5% to almost a hundred thousand in 2020, while face-to-face consultations plummeted from 150,000 to 22,000. This demonstrates the effectiveness and growing acceptance of remote care. Telemedicine has played a crucial role in freeing up hospital space during the pandemic.

COVID-19 positive patients were able to receive follow-up care through secure online platforms, enabling them to be discharged to self-isolation. This shift not only alleviated the strain on hospitals but also showcased the potential of telemedicine in shifting the point of care from hospitals to wherever the patient is located.

Dr. Mesko highlights that the digital revolution in healthcare is not solely about technology but also involves a cultural transformation. The pandemic has pushed an overburdened healthcare system to take the first steps towards more efficient healthcare systems utilizing AI and digital health innovations. Medical professionals were faced with the choice of either providing care remotely or not providing care at all. This stark reality emphasized the need to reevaluate traditional healthcare practices and embrace digital solutions.

In 2013, Sensely developed an innovative virtual health assistant named Molly. Molly is an AI-powered avatar that provides personalized interactions with patients, offering support and gathering vital health information. Patients can engage with Molly on a daily basis, and even multiple times a day if needed. This virtual assistant asks patients about their condition, collecting essential data such as weight, blood pressure, and glucose levels. Based on the patient's specific health condition, Molly asks relevant questions to gain a comprehensive understanding of their well-being.

The seamless transmission of this information back to the patient's healthcare provider is a key feature of Molly's capabilities. This allows healthcare providers to stay informed and be alerted if a patient may be experiencing any health issues. Molly's ability to engage in conversations with patients enhances patient engagement, providing a comforting and supportive experience.

Patients have expressed their preference for interacting with Molly over using traditional health apps on their phones, as Molly's presence is akin to having someone hold their hand during their healthcare journey.

An illustrative video I watched showcases Molly in action, demonstrating the value it brings to patient care. In the video, Molly checks a patient's weight, pulse, and blood pressure, initiating a dialogue about their current condition. Noticing a slightly elevated blood pressure reading, Molly suggests that the patient reach out to a nurse to potentially make adjustments to their clinical routine. Despite being outside of regular office hours, Molly offers to connect the patient with someone who can assist. The patient agrees, and Molly takes the initiative to arrange for a nurse to call the patient back on the next business day. The interactive nature of Molly's interactions and its ability to gather real-time health data greatly contribute to patient engagement and adherence to health-related tasks. Patients find comfort and reassurance in Molly's presence, appreciating the personalized support it provides. This increased engagement level fosters a proactive approach to maintaining good health and empowers patients to take control of their well-being.

Molly serves as an exemplar of how AI-powered virtual health assistants can revolutionize the patient experience. By combining advanced technology with empathetic and personalized interactions, virtual health assistants like Molly have the potential to improve patient outcomes and create a more patient-centered healthcare environment. The utilization of such innovative solutions further highlights the transformative role of AI in healthcare, ensuring that patients receive the support and care they need, even beyond traditional office hours.

Artificial Intelligence in creative arts

AI-generated artwork is on the rise and poised to revolutionize the lives of creative professionals and photographers. This emerging form of visual art is created using artificial intelligence algorithms, allowing individuals to generate stunning pieces of art, illustrations, and even photography based on prompts generated through machine learning. By leveraging pre-existing data and analyzing a vast number of visual arts, AI-generated artwork brings forth new and unique creations. One of the most remarkable aspects of AI-generated artwork is its potential to democratize the art world.

Traditionally, art has been seen as an exclusive realm, requiring extensive training and education. However, with AI-generated artwork, individuals without formal art backgrounds can now express their creativity and produce beautiful pieces of art. This technology breaks down barriers and empowers anyone with an idea or vision to bring their artistic concepts to life.

As a fashion entrepreneur, I understand the immense value of captivating content for a brand, whether through photography or videography. In the past, substantial resources were allocated to photography, graphic design, and copywriting. However, with the advent of generative artificial intelligence, many of these tasks have been streamlined and made more accessible. While significant projects may still require professional artists, smaller-scale projects can now be handled in-house using apps like Photoshop AI for image manipulation and garment design, Midjourney and DALLE for generating ideas and graphics, and ChatGpt 4 for copywriting and content scripting.

This technological advancement has not only saved time but also reduced costs for businesses like mine. The impact of AI-generated artwork extends beyond digital realms and is transforming traditional art forms such as drawing, painting, and photography. AI-produced art offers a quicker and more cost-effective alternative, challenging the traditional approaches. As a result, artists using traditional mediums must now navigate the landscape and find ways to adapt and experiment in order to maintain a competitive edge while embracing the possibilities that AI-generated artwork presents.

AI and machine learning have been making their mark in the creative music industry for some time now. Early examples of AI and machine learning in music include Spotify's data-driven playlists and Shazam, which identifies songs playing in the background. As AI continues to be adopted, it is reshaping the music-making landscape by offering new tools that assist producers, composers, and musicians in creating more sophisticated and personalized works. Voice cloning is one area where AI has gained widespread use in the music industry. This technology utilizes deep learning algorithms to analyze and replicate the nuances of a human voice. For example, the Resemble AI software platform allows for the cloning of famous voices.

In April 2023, a voice-generated song featuring the voices of Drake and The Weeknd caused a stir on TikTok. A TikToker, known as 'Ghostwriter977,' used a voice cloning platform to recreate the voices of these artists. By training the synthetic replica voices using audio files of their songs, the software platform was able to produce synthetic voices rapping the lyrics when the ghostwriter typed or spoke them.

This process, known as text-to-speech or speech-to-speech, showcases the potential of AI in replicating and manipulating voices. This also presents opportunities for streamlining music production. Artists can outsource vocal recordings while on tour by utilizing their voice clones, eliminating the need to return to the studio each time. This can result in significant time and cost savings for musicians and has the potential to revolutionize the industry.

Composition is another area where AI is making waves. Companies like Amper and AIVA have developed AI tools that create compositions for various media, including content, film, and gaming. These applications allow users to mix and match instruments and AI-generated compositions to achieve their desired sound. Users have creative freedom and can download the generated content for commercial or personal use. For example, producers can tweak AI-powered compositions downloaded from the app to fit their music projects, expanding their beat-building capabilities.

AI-powered instruments and plugins are also transforming the music-making process. Google Labs' Magenta offers free plugins to Ableton users that generate melodies, rhythms, grooves, and innovative sounds within the digital audio workstation's session view. Other tools, such as the Orb Producer pack, enable producers to generate basslines, melodies, and sounds using wavetable synthesizers, all seamlessly integrated into their creative process.

Machine learning is even making its way into the mixing and mastering process. Izotope's Neutron and Ozone suites, powered by AI, have gained industry acclaim for their advanced mixing and mastering capabilities. Indie artist communities also utilize AI for machine learning-based mastering through apps like LANDR.

Artificial Intelligence in Web Development

Artificial Intelligence (AI) is a transformative force in the realm of web development, offering automation and problem-solving capabilities that are reshaping the industry. Tasks that once necessitated human intervention, such as coding, crafting website copy, performing SEO tasks, creating graphic design images, and troubleshooting performance issues, are now being accomplished with AI-powered algorithms. These algorithms generate code and automate significant portions of web development work, ensuring accuracy and reducing the need for extensive testing and revisions.

AI's role in optimizing websites for better visibility and performance is instrumental. It can comprehend search engine algorithms, analyze large datasets, and identify patterns to enhance website content and improve search engine rankings. By leveraging AI, web developers can create personalized user experiences through features like AI-powered chatbots and predictive product recommendations. These interactive elements enhance user engagement without requiring additional human effort.

In addition to these applications, AI is making significant strides in User Experience (UX) Design, Web Analytics, and Cybersecurity. AI can analyze user behavior and preferences to create designs that are more likely to engage and satisfy users. Tools like Adobe's Sensei leverage AI to automate complex tasks and enhance creativity, thereby improving the overall design process. Furthermore, AI can analyze vast amounts of data from website users to provide insights about their behavior, preferences, and patterns. This can help businesses to make data-driven decisions and improve their strategies.

In the realm of cybersecurity, AI enhances the security of websites by detecting and responding to threats and attacks in real-time. It can identify patterns and anomalies that might indicate a security breach, allowing for quicker response times.

Looking towards the future, the potential applications of AI in web development are vast. More advanced personalization, improved accessibility features, and the potential for AI to take on more complex tasks are just a few possibilities. However, while the benefits of AI are substantial, it's crucial to consider the ethical implications and potential drawbacks of its use in web development. Issues around data privacy, the potential for job displacement, and the importance of human oversight in AI applications are all important considerations.

Artificial Intelligence in Education

Artificial intelligence (AI) is making a significant impact on education, with China being at the forefront of utilizing AI technology in classrooms. In China, brainwave sensing gadgets are being used to measure students' concentration levels. These devices, made in China, have electrodes that pick up electrical signals sent by neurons in the brain. The neural data is then sent in real-time to the teacher's computer, allowing them to see which students are paying attention and who may be struggling. Reports detailing each student's concentration level at 10-minute intervals are sent to parent chat groups, enabling parents to stay informed about their children's engagement in the classroom.

Teachers in China have noted that the use of these headbands has led to increased discipline among students. Students are paying better attention during class, studying harder, and achieving higher scores.

This technology provides teachers with greater insights into students' concentration levels and helps them tailor lessons to keep students engaged. Overall, students are responding positively to the use of this technology, indicating its usefulness and effectiveness in promoting learning and engagement.

AI has also made significant advancements in personalized learning. Traditional classrooms often struggle to cater to the individual needs and pace of each student. AI-powered e-learning tools enable personalized learning experiences, where students can learn at their own pace, and course content is customized to their individual needs. AI can provide personalized guidance, feedback, and suggestions for teaching materials based on each student's progress, achievements, and strengths.

Grading assignments and assessments can also be automated using AI, reducing subjectivity and bias while providing immediate results. For example, Quizlet, a study platform, uses AI to recommend quizzes and study content based on students' progress and interests, while Carnegie Learning's Mika software adapts to individual student needs, offering additional support for struggling students. AI is transforming the assessment process by automating the grading of papers, essays, tests, and exams. Machine learning algorithms analyze student work, providing immediate feedback and identifying areas of strength and weakness. This allows both teachers and students to tailor their study plans to optimize performance. Companies like Kritik use AI to provide students with feedback and teachers with insights on student performance, facilitating decision-making and better learning outcomes.

In the field of educational research, AI-generated data provides valuable insights. AI tools generate vast datasets that can be analyzed to identify problems and propose solutions. Educational publishers like Pearson are utilizing AI to analyze student performance data and predict areas of difficulty, helping teachers make adjustments to improve learning outcomes.

CHAPTER FOUR

THE MAGIC OF HUMAN INGENUITY & CREATIVITY

In previous chapters, I explored how various work sectors have started integrating Artificial Intelligence to increase productivity, reducing job opportunities for humans. However, despite AI's advantages, it has limitations. AI can replicate many human tasks but cannot replace human creativity and originality in the workplace.

Humans possess an elemental creative spark that resonates emotionally and attracts the world. This relies on human originality and innovation, highlighting the irreplaceable human role in delivering unique crafts and ideas. Therefore, it is unlikely AI will wholly replace humans in the foreseeable future. AI identifies and utilizes pattern similarities to make decisions. When trends change narratives, human intelligence and creativity become necessary to reorient things.

For example, Dr. Bertalan Mesko's "trolley problem" thought experiment illustrates this AI limitation. Imagine a runaway trolley barreling toward five helpless workers. You stand near a switch that will divert the trolley to another track, saving the five people. However, that track has one worker who would then be killed. Doing nothing means the trolley kills five people. Pulling the switch means actively killing one person to save five. Which choice is more ethical? Would you kill one person to save five? With AI often controlling such decisions nowadays, we are all metaphorically tied to the tracks.

This classic ethical dilemma, called the trolley problem, was created in 1967 to evaluate the moral implications of actions based on outcomes. Now, with the rise of self-driving cars, it exemplifies the challenges of AI decision-making. Imagine your self-driving car is boxed in by traffic when a heavy object falls from a truck ahead. Your car cannot stop in time and must decide - hit the object head-on, swerve left into an SUV, or swerve right into a motorcycle. As self-driving cars become more prevalent, such scenarios will inevitably arise. So how should the AI decide? Should it panic and prioritize your safety at all costs like a human might, even if it causes more casualties? Or should it logically evaluate and choose the outcome with the fewest projected deaths, essentially premeditating your homicide? There are no easy answers.

Moreover, what if the AI weighed not just potential casualties, but the calculated value of the lives spared? Human intelligence and creativity have irresistible potential in the AI age. We must harness their power for continued innovation and progress.

Understanding the Basics of Human Intelligence and Creativity

Defining human intelligence is a bit like trying to describe the scent of a rose. It's a beautifully intricate concept, one that holds a special place in our hearts as humans. At its most basic, intelligence is our magic wand for problem-solving, our guide through the wilderness of life. It helps us find food and shelter, choose our life partners, and dodge the metaphorical bullets life sometimes fires our way. But intelligence isn't a one-trick pony; it's a magic show full of incredible tricks such as acquiring knowledge, learning, creativity, strategic thinking, and critical judgment.

Imagine a Swiss Army Knife, a tool with multiple gadgets, each with its own purpose and priority. That's how you can think of intelligence a versatile toolbox ready for whatever life throws our way. The fundamental tools within this toolbox are the abilities to gather, store, and apply information. We perceive the world through our senses sight, hearing, taste, touch, smell and use this sensory information to navigate and respond to our surroundings. Monitoring our internal states, like feeling hungry or tired, is also a part of this process. These signals help us understand our body's needs and guide us in fulfilling them, like searching for food when we're hungry or resting when we're tired.

Storing information, also known as memory, is another indispensable gadget in our intelligence toolbox. Our ability to remember and recall information enables us to learn from our past, navigate our present, and anticipate our future. Imagine if each time you encountered a situation, it was as if it was happening for the first time. It would be nearly impossible to learn or adapt. Memories, which could be of events, locations, connections, or actions, form our understanding of the world and shape how we act and react.

Learning is yet another critical tool in our intelligence kit. It's like assembling a complex jigsaw puzzle, one thought or action piece at a time. Through repetition and practice, we learn new skills and behaviors, tweaking and modifying them to fit different situations. Our ability to alter our behavior based on past experiences, thereby enhancing our problem-solving capabilities, is what makes learning a dynamic and vital part of intelligence.

Even seemingly simple creatures can display incredibly intelligent behavior through these basic tools of intelligence. For instance, an episode by Kurzgesagt titled "What Is Intelligence? Where Does It Begin?" highlighted the surprising behavior of an acellular slime mold, a single-celled organism.

Despite its simplicity, this slime mold explores its environment, leaves a slime trail as a memory marker, and adapts its behavior to save energy demonstrating memory and learning at their most fundamental.

As problems grow more intricate, our intelligence toolbox expands and becomes more sophisticated. More complex animals possess a diverse "Library of Knowledge", storing a range of associations, relationships, and tricks that allow them to handle a greater variety of problems. Raccoons, with their notorious love for human food, are a prime example. They've developed a suite of theoretical and practical skills that enable them to navigate human-made structures and objects. Their ability to pick locks, open windows, and devise other strategies to procure food showcases an advanced use of the intelligence toolbox.

Among the most impressive gadgets in our intelligence toolbox is creativity. It's not just about creating beautiful art or music, it's about innovative problem-solving. It's about looking at seemingly unrelated things and creating something new and valuable, a process that requires making unique and unconventional connections.

Finding unexpected answers to difficulties is a common aspect of creativity. For example, instead of throwing pebbles into a tub to raise the water level, a guy tips the tub over faster, which is a more creative solution. This demonstrates his ability to think creatively, applying skills and knowledge in new ways. Creativity also frequently entails the clever use of tools. Across cultures, we see people innovatively employing available resources. From indigenous tribespeople using sticks to reach honey in trees, to sailors assembling floating devices from debris to stay afloat, the creative use of materials illuminates the boundless potential within the human intelligence toolbox.

Legend has it that when guitarist Keith Richards couldn't find a guitar pick, he used a knife instead, inadvertently creating the iconic riff for the Rolling Stones hit "Satisfaction." This exemplifies how creativity thrives under pressure and resourcefulness. By imagining possibilities beyond an object's intended purpose, we spark ingenuity.

The Apollo 13 astronauts, faced with a life-threatening equipment failure, miraculously engineered an improvised carbon dioxide filter from items like duct tape, socks, and plastic bags on their ship. Their creative solution saved their lives and the whole mission. Even in the most challenging situations, human intelligence finds a way forward. Our minds hold seemingly infinite possibilities. By fostering creativity and imagination, we can uncover innovative solutions to the most daunting challenges. The toolkit of the intellect remains our most valuable asset.

The myth of 'being intelligent

Allow me to take you on a journey, one that starts in my birthplace in the southern part of Nigeria, in a state called Edo. This place is just about a four-hour drive, circumstances permitting, from a renowned place known as Ekiti. Ekiti is famous, or infamous depending on who you ask, for boasting the highest number of professors in Nigeria. This reputation has fostered a widespread rumor, a belief so pervasive that many have come to consider Ekiti as a bastion of intelligence.

My early impressions of intelligence were heavily influenced by this belief system. In my world, intelligence was directly and unequivocally associated with academic achievement. More specifically, proficiency in certain subjects such as mathematics and the sciences were upheld as the gold standard.

If you could solve complex equations or understand the mysteries of biology, you were crowned 'intelligent'.

To illustrate this point, let's journey back in time to a day in middle school. I arrived home clutching my school report card, my heartbeat drumming in anticipation. As was customary, my father's gaze would always first gravitate towards my mathematics score. This time was no different. His eyes scanned the page and found what they were looking for - a mathematics score that left much to be desired. His gaze then shifted to the teacher's remark - "He needs to pay more attention in class". A shake of his head and a disappointed sigh, and he was done. My stellar grades in fine art and home economics overlooked, ignored, and rendered insignificant. My intelligence, my worth as a student, were measured solely by my performance in mathematics. This narrow perception of intelligence wasn't unique to my father. Many of my friends experienced the same treatment. Looking back, I can empathize with my father's viewpoint, and it's not hard to see why he valued certain skills over others.

You see, my father was a product of his time. Born in the fifties, a time following the industrial revolution, he was molded by a structured system that placed high value on formal education and certain types of knowledge. Subjects such as mathematics, physics, and other science-oriented courses were fundamental tools needed to thrive in a world shaped by the industrial revolution. Tragically, he lost his father, my grandfather, while he was just a teenager in middle school. Without a family safety net or inheritance to rely on, my father had to pay a heavy price to rise from these circumstances.

He fought his way through, earning a degree and landing a job as an accountant, and eventually becoming a manager at the then newly established Nigeria Bank. He was, in every sense, a product of a system that cherished certain skills and disregarded others. His life was an embodiment of that time, and his beliefs, a reflection of the societal values he grew up with and the Myth where academic success is tied to intelligence.

As I weave this tale, my hope is to not just recount my experiences but also to illuminate the need for a broader understanding of intelligence. To understand intelligence as a multifaceted trait that isn't confined to a few subjects or skills. To appreciate the diversity of our cognitive abilities and to celebrate all forms of intelligence. To redefine what it means to be 'intelligent'.

In his book "The Element," Ken Robinson makes a profound point about the relationship between intelligence and creativity. He mentions that when he asks people to rate their intelligence and creativity on a scale of 1 to 10, most rate themselves as average on both. However, two-thirds to three-quarters of people give themselves different scores for intelligence and creativity. Robinson contends that intellect and creativity are inextricably linked and that creativity is a component of intelligence. We often underestimate the intelligence of creative people.

For example, I had a childhood friend named John who could draw amazingly detailed comic book heroes like Batman and Spiderman. We were in awe of his artistic talent, waiting patiently as he carefully sharpened his pencils before sketching. But John struggled academically and had a reputation for not being bright. If only we had understood intelligence more deeply back, then. People often equate intelligence solely with logical and mathematical skills. But in reality, intelligence encompasses a broad range of abilities, including creativity, spatial reasoning, interpersonal skills, and emotional intelligence.

Recognizing the diverse manifestations of intelligence allows us to appreciate the intellectual gifts of creative individuals like my talented friend John.

Debunking the Myth

The origins of intelligence quotient (IQ) testing trace back to the Industrial Revolution's narrow definitions of intelligence as solely verbal and mathematical reasoning. This limited scope fails to encompass the rich diversity of human intelligence. As Ken Robinson describes, intelligence manifests in myriad forms like analytical, creative, and practical aptitudes. Valid criticisms exist of defining intelligence based solely on IQ tests.

Howard Gardner's influential theory of multiple intelligences proposes that intelligence includes linguistic, musical, logical-mathematical, spatial, bodily, kinesthetic, interpersonal, and intrapersonal capacities. These intelligences are distinct and independent dimensions.

I firmly believe each person possesses unique gifts and talents constituting their intelligence profile. Intelligence differs substantially from one individual to the next, with people applying their abilities in varied ways. Some discover their proclivities early in childhood, while others explore diverse pursuits before embracing their strengths. We should nurture our own intelligence rather than conforming to rigid definitions.

Creativity is the key; it entails applying intelligence to generate value. While academic scholars like Albert Einstein leveraged creativity remarkably within their fields, creative geniuses like Michael Jackson revolutionized music and entertainment, also immensely valuable industries. It would be misguided to suggest Michael lacked intelligence simply because he was not an academic.

The story of Hollywood superstar Tom Cruise further debunks the myth that academic intelligence alone determines success. Cruise struggled tremendously in school due to dyslexia, a learning disability making reading extremely difficult. Despite perseverance, he scored poorly on assignments and tests, unable to demonstrate his knowledge. Teachers dismissed Cruise as slow and unintelligent.

This constant failure devastated his confidence, making him feel inadequate. In high school, Cruise regained some optimism by thriving in sports and discovering a passion for acting. While dyslexia still challenged him, debate and drama allowed Cruise to shine creatively and interactively. Acting gave purpose to the boy deemed unintelligent academically. After graduation, Cruise pursued his acting dreams in New York. However, dyslexia posed immense obstacles to memorizing scripts and lines. After rejections, Cruise got his big break in Risky Business. His evident talent and charisma opened doors, leading to stardom with Top Gun.

Cruise's story illustrates that intelligence alone does not determine success. Despite dyslexia, through dedication Cruise became a global superstar by embracing his gifts rather than limitations. His journey should inspire us all to pursue our dreams without constraints based on notions of innate talent or intelligence. With creativity and drive, we can defy expectations, as Cruise did.

Human Creativity

As stated earlier, creative tendencies are multifaceted and unique to each individual. Throughout history, creativity has been essential to human advancement, innovation, and self-expression. For thousands of years, humans have displayed remarkable creative abilities, from prehistoric cave art to modern technologies. Still, the question of why humans are creative has puzzled many researchers.

Recent studies provide insightful explanations that humans' brains evolved to encourage creativity through a reward mechanism, and creativity is innate to our genes. The brain's structure and creativity reward system explain why creative activities intrinsically motivate and satisfy us. Creativity is distinctive to humans and has enabled the invention of tools, techniques, and ideas furthering civilization. Experiments by Drexel University researchers revealed creative breakthroughs activate the same brain reward centers as delicious food, addictive substances, and sex. This suggests our brains evolved to reward creativity, driving human innovation and discovery. These findings support creativity's crucial role in shaping cultures, economies, sciences, arts, and beyond.

Neuroscience has also revealed unique brain characteristics that foster creativity. The frontal lobes, in particular, are associated with creative thinking. Evolutionary alterations in brain architecture contributed to the distinctive developments that distinguish humanity from other intelligent creatures. It is widely accepted humans are born creative. Our innate curiosity and drive to explore various forms of expression have progressively shaped society. Even very young children exhibit incredible creative energy, whether painting, building sandcastles, or inventing games.

Evidence indicates this creative drive is hardwired from a young age, making creativity an inherent human trait.

The idea that human creativity springs from being fashioned in the likeness of a creative God is a captivating one. Many of the world's major religions hold at their core the belief that humans are a reflection of God's image, embodying divine qualities, one of which is creativity. This viewpoint provides a deep understanding of where our creative impulses come from.

Let's take a moment to delve into the biblical account of creation. Genesis 1:1 paints a vivid picture: "In the beginning God created the heavens and the earth." This suggests that the universe itself is a product of God's boundless imagination and innovative prowess. Fast forward to Genesis 1:27, and we find that on the sixth day, humans were crafted in God's very likeness. This creation was not just a physical one but also an endowment of attributes, including the ability to create. By the time we reach Genesis 1:31, there's a confirmation of the excellence of God's creative works. And in Genesis 2:2, after bestowing upon humans the gift of creativity, God takes a well-deserved rest.

But what does this mean for us? Our creative potential isn't limited to just art. It spans a vast spectrum, from groundbreaking ideas and inventions to solutions that can transform the world for the better. When we harness our inherent creative abilities, we're essentially living out our spiritual purpose, mirroring God's own creative essence.

Christianity, for instance, has always celebrated creative expression. Music, dance, visual arts – they're all woven into the fabric of worship, fostering a deep sense of connection with the divine. When we view creativity as a gift, a spark of the divine within us, it becomes a powerful force. It's not just about personal fulfillment; it's a spiritual journey.

In recognizing and embracing our creativity, we're not just tapping into our personal potential. We're connecting with something much larger, a divine legacy. By fully realizing this gift, we open up a world of possibilities, not just for ourselves but for our communities and the world at large. So, the next time you create, remember: it's a reflection of the divine within you.

Human creativity manifests in remarkably diverse ways, from breakthroughs in science and technology, to cultural works that inspire and connect society, to everyday problem-solving and resourcefulness. Our shared creative nature is a unifying tie, reflecting the divine creative spark present within all humanity. Appreciating this spiritual dimension expands our understanding of creativity's origins, significance, and boundless potential when aligned to righteous purposes.

Embracing the element

If you've read this far, you've likely spent time reflecting on your own innate strengths and where your intelligence truly comes alive. It's so important to identify those activities that spark your creativity and ignite your talents. When you know what energizes your intellect, you can embrace your uniqueness and become irreplaceable.

The renowned creativity expert Sir Ken Robinson refers to this sweet spot as finding your "Element." He describes the Element as the meeting point between natural aptitude and personal passion. It's where your innate talents intersect with deep intrinsic motivation.

When those two forces combine, your creative potential can be unleashed in extraordinary ways. That is when our innate talents combine with deep intrinsic motivation, the sparks of inspiration can ignite into world-changing creativity. On one hand, aptitude refers to the raw potential we are born with. This includes abilities like imagination, aesthetic judgment, and talents in specific domains like mathematics, music, or design.

These cognitive tools provide the basic mental hardware needed for creative work. Just as a master sculptor requires fine motor skills, creativity draws on particular genetically shaped skills. However, raw talent alone does not make a creative genius. Passion provides the fuel that drives the exploration of new ideas. Those with a burning, innate curiosity and drive to master a skill or subject area can overcome even modest innate gifts. Passionate energy compels dedicated practice, exploration of novel angles, and the perseverance to see creative projects through.

While aptitude is largely fixed, passion can also be nurtured through exposure, training, and self-discovery. Robinson argues that the pinnacle of human creativity occurs when exceptional natural abilities are paired with an enduring internal zeal. When talent and passion converge, they feed each other's growth in a virtuous cycle. Interest drives the gifted to dedicate thousands of hours to mastering their field, while their accomplishments validate and reinforce their passion. Even for the talented, falling out of love with a subject often marks the decline of their creativity.

So within us all lies a creative spark waiting to be ignited. The path is to discover those subjects and problems where our innate gifts and intrinsic passions intersect. By listening to our talents and our muse, we can find the kindling needed to fuel breakthrough innovations and creative mastery. For at the meeting point between aptitude and passion, we unlock the most extraordinary capacities of the human mind which we can call the 'Element".

Every so often, we encounter individuals who have truly found their life's purpose, their "element," manifesting as an unparalleled mastery of their craft. My journey back to academia in 2022 for a Master's degree in education brought me face to face with one such beacon of inspiration: Rania.

After several years of focusing on my entrepreneurial ventures, I had reservations about reentering the academic realm. Anticipating challenges in readjusting to scholarly pursuits, it was Rania's teaching methods that eased my transition. Instead of the usual academic rigidity, her classroom radiated warmth and connection. In an astounding act of dedication, she took the time to learn and remember each name of her 80-strong student roster. With Rania, there was a sense of community; every student mattered.

Describing Rania's teaching approach is akin to detailing an exhilarating voyage through space. Every class, though intellectually invigorating, left one both satisfied and eagerly awaiting the next session. But what is more astounding is that teaching was never Rania's first choice. Rania's journey into the world of academia began not out of ambition but from a place of love for learning. As a young 18-year-old university student, simultaneously navigating the waters of early marriage and impending motherhood, her educational pursuits were momentarily halted. But her passion for knowledge led her back, this time through an open enrollment program, striking a balance between her roles as a mother and a student. During this period, the profound impact of her professors - who treated students as intellectual peers rather than subordinates - set her on a path she hadn't foreseen.

Starting off as a part-time teacher, Rania's dedication and prowess soon elevated her to the role of a MA program director at the university. For her, teaching wasn't just about transferring knowledge; it was about creating a universe of discovery. "In that classroom," she would say, "everything outside disappears, and we embark on passionate book journeys together."

Like many, Rania faced hurdles in her early teaching career. From managing diverse classrooms to crafting engaging lessons, the responsibilities seemed daunting. Yet, she stood firm, evolving and refining her approach over time. Instead of seeing students through the narrow lens of academic performance, Rania viewed them as intellectual beings on their own unique journeys. Her ethos centered on nurturing minds, kindling curiosity, and leading students to those transformative "Aha!" moments.

Rania's vision for her classroom is clear: a space of genuine human connection, where every student feels seen, understood, and inspired. It's a space where exploration is encouraged, and new realms of understanding await. Through her story, we are reminded of the beauty of discovering one's true calling, even when it's not part of the original plan.

In a world rapidly advancing towards automation and artificial intelligence, individuals like Rania serve as a testament to the irreplaceable magic of human connection. Machines might replicate knowledge, but they can't emulate the passion, creativity, and profound impact educators like Rania have on their students.

She is the embodiment of the belief that when you're truly in your element, even the most advanced AI pales in comparison.

There are many people thriving in their element across different fields. These individuals harnessing their innate talents and passions are not necessarily famous celebrities, but they are irreplaceable because of the immense value they bring through their uniqueness. While their names may not be on the "wall of fame," these people are alive in their purpose every day. What they have in common is discovering work at the intersection of their natural aptitudes and intrinsic motivations. They exemplify the kind of creativity, innovation, and inspiration that arises when following one's calling.

There are countless stories waiting to be told of everyday people who have found their element and channeled their gifts to uplift their communities and workplaces. Whether teachers igniting young minds, engineers solving complex problems, or artists moving audiences emotionally, these unsung heroes often make the greatest impact through their originality and dedication. They thrive by embracing their individuality rather than conforming to preconceived notions of success or talent. Their fulfillment comes from aligning work with their values, passions, and creativity. This inner alignment allows them to achieve greatness and meaning on their own terms.

These purpose-driven individuals represent the irreplaceable human spirit. Their diverse expressions of excellence and dedication show why artificial intelligence can never replicate the full spectrum of human potential. When people follow their element, they unlock reserves of creativity, empathy, and wisdom that technology lacks.

CHAPTER FIVE

THE IRREPLACEABLE HUMAN TOUCH

Givers Win Hearts

As our world evolves, harnessing our unique gifts to meaningfully connect with others becomes increasingly vital. Our value depends on how we serve people and sustain connections. Interpersonal skills have always been crucial, enabling both good and evil depending on application. For instance, Adolf Hitler exploited his charismatic ability to connect with millions for malign ends, epitomizing a "taker" mentality. Abraham Lincoln, one of history's most beloved presidents, embodied the "giver" spirit focused on the greater good.

In his book "Give and Take," Adam Grant presents a framework distinguishing givers, who contribute more than they receive, from takers who prioritize self-interest, and matchers who strive for equal give-and-take. Of these three, givers are indispensable in teams and organizations. Their willingness to give more accelerates progress through generosity. Upon taking office, Lincoln appointed past rivals to cabinet positions, valuing their potential contributions over personal grudges. This giving mentality earned immense respect, with Alexis de Tocqueville praising Lincoln for holding "the most perfect empire ever won by the genius of common humanity."

My lecturer Rania shared an illustrative story from when she was starting out as an assistant teacher. She was so passionate about teaching that when the main teacher she had been assisting departed, the head of the university approached Rania with an offer. He asked her to take over teaching the course full-time. As Rania recounted, the head said enthusiastically, "I will pay you to teach full-time!" Rania was amazed and thrilled by this opportunity. In her mind, she felt that she would gladly pay the school for the privilege of teaching full-time, since it was something she loved doing. She had such a deep drive to share knowledge and empower young minds.

This story truly encapsulates the mindset of a giver - someone who is so intrinsically motivated by their work that they would almost do it for free. For givers like Rania, contributing their skills and expertise is a reward in itself. Their passion comes through in the sheer joy and fulfillment they feel when giving to others. Rania's tale illustrates how givers are driven by purpose over paychecks. Her reaction to essentially being asked to get paid for following her calling highlights the source of energy and vitality that fuels givers. Rather than being depleted by constant giving, they are energized by opportunities to uplift others.

Today, interconnectedness enables givers to expand their positive influence. As collaboration across domains grows, individual accomplishments matter less than enabling team success. Givers freely share knowledge, take on unpopular tasks, and provide help without expecting anything in return, building trust and cooperation fueling collective achievement. Though automation threatens many jobs, givers are unlikely to be displaced because of their irreplaceable human touch. Work structures now favor giver qualities in many fields.

As the service sector expands, jobs involve more collaboration. Healthcare, retail, hospitality all rely heavily on caring service to others. Even in technical fields, collaboration tools make sharing knowledge easier. This interdependence allows givers to be exceptionally productive, as their desire to serve aligns with required tasks. While self-interest remains, giving has become a respected principle in most modern organizations. Books like Grant's and Simon Sinek's "Leaders Eat Last" show how an others-first approach achieves greater loyalty and performance. Though dominance tempts some, most recognize givers enact the most enduring positive influence. Their choice to uplift others through human connection remains impossible to replicate artificially. The beautiful act of contributing to others remains quintessentially human.

Technology Cannot Replicate Empathy, Compassion, and Human Connection

Human relationships fulfill our profound spiritual need for meaning and purpose. When we cultivate bonds through mutual care, sacrifice, and service, we experience life as deeply meaningful. Serving others' needs before our own provides a sense of purpose beyond the self. Simple acts of generosity elicit a transcendent feeling of human connection. By walking together through hardship and triumph, sharing vulnerability and trust, relationships allow us to embark on sacred journeys of intertwined growth. Our souls evolve through the compassion, forgiveness, courage, loyalty and sacrifice that love inspires. Relationships shape life's meaning.

In his book "Emotional Intelligence," Daniel Goleman demonstrates how empathy is key to meaningful relationships and success. Empathy, the ability to deeply understand and relate to how others feel, enables bonds far richer than any technology can emulate. While technology allows efficient communication, it strips away the nuanced emotions distinguishing human relating. Texts, emojis and chatbots, while useful, distance us from the complexity of face-to-face interaction. Devoid of empathy, they cannot comprehend the full depths of joy, grief, fear and hope defining the human experience.

I previously contacted PayPal help late at night regarding an account problem. An automated voice guided me through instructions, including asking me to briefly describe the situation. This demonstrated technical sophistication. However, I desired to fully express my dissatisfaction and obtain true understanding and an empathetic human reaction. While digital interactions are beneficial, they lack the emotional richness seen in human relationships. There are intangible components of human bonds that technology cannot capture.

Consider healthcare, where technical advances still cannot substitute for the human touch. Doctors' emotional skills cultivate trust and tailor care through empathy. They can comprehend patients' fear, anxiety and suffering to provide comfort. Medicine remains an innately human profession. For all its benefits, technology is limited to pragmatic function. Without spiritual capacity for sacrifice and mutual care, even the most emotionally intelligent AI cannot replicate relationships' profound meaning. It is the essence of being human - the intertwining of souls on journeys of compassion.

Relationships' higher purpose eludes virtual bonds. While technology effectively simulates emotional intelligence, the deepest meaning arises through selfless human bonds. True friendship, family and community shape life's significance.

No machine can experience the sacred interdependence at the heart of human connection. Herein lies the zenith of our humanity.

The choice to apply innovations generously rests with humans

"There is nothing either good or bad, but thinking makes it so." – *William Shakespeare*
This perspective applies aptly to technological innovation. While offering exciting possibilities, humans retain the ethical responsibility in how these tools are applied. The choice to utilize discoveries and inventions for good versus harm ultimately lies with people rather than the technologies themselves. At their core, tools like artificial intelligence and automation are neutral - it is human beings who determine whether to apply them compassionately or destructively. A hammer can build a home or harm others depending on the wielder's intent. Fire can sustain or destroy life based on its use.

The same applies for cutting-edge breakthroughs like gene editing, augmented reality, and robotics. Their immense powers can uplift humanity or deepen injustice depending on the values of those guiding their impact. It is humans who write the algorithms, conceptualize, curate and oversee those algorithms, and produce the desired outcomes. Humans are the ones who have authority over AI, and it is their actions that determine the AI's nature. The fears of humans losing their jobs due to AI are not new. In Ancient times, industrial revolution gave birth to new companies, which opened up new job opportunities. Similarly, the generative AI revolution can create liberation and freedom for the labor force, allowing individuals to explore their unique personal emotional skills that no computer can replicate.

These skills include, but are not limited to, creativity, problem-solving, empathy, and leadership. By developing these skills, people can take advantage of the opportunities that the future job market will offer.

Throughout my life, I have cultivated my passions for fashion, content creation, and graphic design. Over the years, I have amassed a wealth of experiences that I have often contemplated collecting into a book. However, I lacked confidence in my ability to eloquently convey my ideas in writing. I worried that my grammar was inadequate and my self-expression poor. I told myself that I would need the help of a professional writer to refine my thoughts into polished prose. That changed when generative AI like chatGPT emerged.

This technology has been a revelation, providing me a pathway to organize my concepts into coherent, engaging narratives far exceeding my capabilities. With the assistance of AI writing tools, I have finally been able to produce this book, which would have been impossible otherwise. Despite reading extensively and researching diligently, I do not possess the natural writing talent to have developed this manuscript solo. My skills lie in creative pursuits like design, not eloquent writing. Generative AI has proven invaluable in extracting my raw ideas and transforming them into sophisticated, reader-friendly prose. It has helped me work through structural decisions, flesh out details, and find the right tone and flow. While I supplied the original vision, AI enabled me to share my experiences and perspectives in an accessible, compelling way for the first time. I am humbled and grateful for this technology that has empowered me to fulfill my long-held dream of authorship. I firmly believe that everyone has important stories to tell and insights to share if given the proper tools.

My journey shows how emergent AI can expand creative possibilities by helping people organize and articulate their narratives who otherwise would struggle alone. With the continued evolution of these writing assistants, more untold stories will finally find their way into the spotlight.

Technology is simply a means, not an end. It provides capabilities to improve lives only if guided by moral wisdom and human conscience. Technological change inevitably intersects with complex social, political and environmental dilemmas that require nuanced ethical reasoning. While inventions create new realities, humans retain responsibility in steering progress toward justice, sustainability and shared prosperity. We must governance innovation with great care and foresight. Ultimately, technology is only a tool. The choice to employ it generously and for the common good remains a fundamentally human one. Our values must determine whether discovery leads to equitable advancement or compounds inequity and harm. With conscience as our guide, we can be writing the moral code for technological change.

The Best of Both Worlds

In the previous section, I established that technology itself is neither inherently good nor bad, its impacts depend on how humans choose to apply it. This same principle applies to automation and AI. Rather than viewing these technologies as inevitable threats to our livelihoods, we can choose to perceive the opportunities they present if harnessed wisely. Even if certain jobs are lost to automation, we must adapt and find new ways to apply our skills productively in an AI-powered world. This technology is here to stay and will only become more advanced, so the question is how to evolve with it rather than be displaced by it.

In fact, I believe artificial intelligence provides valuable models for discovering our unique element and strengths. Even in everyday life, we have adopted tools that help automate mundane tasks, freeing us up for more meaningful pursuits aligned with our passions. AI can play a similar role on a societal level - taking over repetitive, unfulfilling work to expand human capacity for creativity and innovation.

In his book "Atomic Habits", James Clear discusses how automation, and artificial intelligence can serve as effective tools to remove the friction that comes with bad habits. According to Clear, friction "refers to the resistance that a task creates while you are trying to complete it." By reducing the resistance, tasks become easier to perform, and habits become more effortless to maintain. Clear cites an example of how automation and AI can help individuals achieve their desired habits regarding the use of automation in the airline industry. According to Clear, pilots typically have 7 minutes available to make a decision in case the airplane experiences a crisis during the flight. However, during a crisis, human instincts can sometimes be paralyzed. Pilots can find it hard to read instruments, stay calm, and make life-saving decisions.

To reduce human error and make air travel safer, Clear cites the use of automation. Planes are equipped with advanced sensors and algorithms that help detect potential problems and alert the pilots to take corrective actions. In some cases, automation can even take over control of the plane to maintain its safety. Clear acknowledges that automation is not without its flaws. It can malfunction, fail, or make errors. However, in situations like the ones faced by pilots in-flight, the benefits of automation can be immense.

By integrating automation into the decision-making process, pilots can make more informed decisions with a lower chance of error, effectively improving the safety of the passengers in their care.

As automation continues advancing, thoughtful examination of how human skills can complement machines is crucial for optimal efficiency. The wisdom of Socrates rings true: "an unexamined life is not worth living." Taking stock of our innate strengths and limitations, along with gaining proper AI education, are important first steps. Careful consideration reveals the complementary capabilities of humans and AI. Machines excel at tireless computation, analysis, precision and processing vast datasets - tasks far beyond human cognition. Yet humans supply the contextual reasoning, empathy, ethics and creativity that AI lacks. We can make intuitive connections and imaginative leaps in reasoning that confound machines. Rather than a threat, we must see AI as amplifying human skills when judiciously applied.

For example, a designer could leverage generative AI to iterate designs rapidly, but rely on uniquely human judgment to determine the ideal creative direction. A doctor might use medical AI to analyze patient data and suggest possible diagnoses, while maintaining the irreplaceable human care and bedside manner so vital for healing.

Since AI has no emotional intelligence, distinctly human strengths like communication, teamwork and empathy become even more essential. Truly successful integration will require seamless collaboration between people and technology. While chatbots manage simple customer service queries, human agents supply the nuance, complexity and empathy needed for negotiating tricky interpersonal situations. To complement AI's rise, workers must adopt a growth mindset and commit to lifelong learning, developing skills like creativity, critical thinking, collaboration and complex communication that exceed AI's capabilities. With proper perspective and education, we can create a collaborative future where AI amplifies human potential rather than replacing it. The key is thoughtfully applying our complementary strengths alongside automation to unlock new possibilities.

EDUCATION TO THRIVE IN AN AUTOMATED WORLD

Self-directed learning and continuous upskilling

"The whole purpose of education is to turn mirrors into windows." As Sydney J. Harris eloquently states, the aim of learning should be opening our minds to new worlds that stretch our perspectives.

However, before exploring how to revolutionize formal education, we must acknowledge that the onus of lifelong learning begins with individuals taking initiative for their own continuing education. In a rapidly evolving job landscape, the skills and knowledge acquired through schooling have a limited shelf-life. To stay employable, adults must become self-directed learners, proactively seeking out the training needed to augment their abilities. From online courses on platforms like Udemy, Alison, Coursera, and Google Skillshop to bootcamps, YouTube tutorials, and mentoring, pursuing continual learning outside of traditional classrooms is critical. By taking ownership of upskilling, we begin to transform the mirror of the self into a window onto possibilities for growth.

Authors Erik Brynjolfsson and Andrew McAfee highlight the revolutionary potential of self-directed online learning through Massive Open Online Courses (MOOCs).

They recommend students and workers embrace these platforms to acquire new skills. MOOCs have fundamentally transformed education by removing traditional barriers like geography, tuition costs, and rigid schedules. They allow learners to direct their own education, participating at their own pace. MOOCs provide unlimited enrollment in courses from leading professors at prestigious universities.

This democratizes access to high-quality instruction globally. Learners worldwide can use MOOCs to gain competencies critical for the digital era, from programming to business analytics. Interactive forums and peer-reviewed assignments mirror key benefits of conventional classrooms. Since MOOCs involve zero marginal cost per additional learner, they empower millions by eliminating financial obstacles. However, the process of self-education can be a daunting task, and many people struggle to stay motivated and committed. This is where partnering with like-minded individuals comes in handy.

The scenario described by author James Clear illustrates the effectiveness of partnering with others to engage in self-education. The two individuals in the story, Mattan Griffel and Jameson Detweiler, who wanted to learn how to code but found it difficult to maintain consistency or "find time" in their busy schedules. They decided to form a "team" of two and commit to daily coding lessons, even if they were only small amounts of code. Each day, one of them sent the other a small chunk of code via text message or email, and the other would review it and provide feedback. They held each other accountable to complete their lesson every day, and it worked. Mattan eventually built a successful tech startup, and Jameson created a coding school. Both credit the daily habit of completing small coding lessons with their success.

The rapidly evolving nature of technology means that new skills and knowledge are always in demand, and individuals must now learn how to perform certain tasks themselves rather than outsourcing them. Partnering with like-minded individuals to acquire these skills offers several benefits, including peer support, peer learning, accountability, and feedback that can improve learning outcomes.

In their seminal book "The New Division of Labor," Frank Levy and Richard Murnane make a vigorous case for strong foundational literacy and numeracy. They contend these fundamentals furnish the scaffolding to then attain higher-order cognitive and noncognitive abilities like critical thinking, ideation, and complex communication. In their view, academic building blocks allow deeper, transferable learning.

Alternatively, tech experts Erik Brynjolfsson and Andrew McAfee argue that succeeding in the digital economy requires capabilities beyond conventional literacies. They highlight skills like large-frame pattern recognition, intuition, and multidimensional communication. Importantly, they believe student-driven, passion-fueled learning best cultivates these skills, not top-down instruction. Memorization and passive learning fail to nurture creativity. Is this a binary choice between structured fundamentals and unstructured human-centric skills? Likely not, the optimal path forward balances both approaches. Solid literacy and numeracy do enable accessing specialized knowledge to apply creatively. Self-directed learning also plays a pivotal role in nourishing imagination and ingenuity. The solution may reside in blending rigorous foundations with ample space for inspired exploration.

Education could provide broad academic grounding through structured learning focused on literacies and disciplines like science, history, and arts. This establishes bases for growth. However, conventional instruction must also make room for creativity and student-driven projects tailored to interests. Passion powers engagement and purpose. Assessments should evolve similarly. Alongside academic testing, evaluating soft skills growth through portfolios of creative work and community impact data better captures human talents. Hybrid human-AI classrooms can combine structured knowledge delivery with personalized enrichment and remediation.

In essence, dichotomizing education between fundamentals and creativity is counterproductive. We need both. The task ahead is thoughtfully integrating formal learning with opportunities to cultivate humanity's creative spirit. This fusion promises to unlock potentials within every student to meet the challenges ahead.

Revolutionizing formal education

Early childhood represents the most pivotal stage for igniting a lifelong love of learning. The quality of education we provide to children in their formative years shapes the adults they will become.

As Sydney J. Harris highlights, education must expand perspectives, not limit potential. With emerging technologies like AI disrupting traditional paradigms, determining the ideal approach to early childhood education is critical. We must get it right from the start.

According to scholar Dr. Glenda Mac Naughton in her insightful book "Shaping Early Childhood," there are three models for early education: conforming, reforming, and transforming.

The conforming approach involves highly prescribed, standardized education where learners must fit into set categories and meet predefined outcomes. This model lacks flexibility to accommodate diverse individual needs. The conforming approach utilizes a rigid, "one-size-fits-all" framework that does not adapt to students' unique abilities, backgrounds, strengths and challenges. Learners are expected to conform to predetermined curricula and methods of assessment with little personalization or deviation. This approach draws heavily from prevailing knowledge frameworks and best practices at the time, often failing to consider new paradigms or long-term solutions to build student capacity. The focus is on meeting standardized targets and benchmarks through established techniques, not reimagining learning environments.

Conforming models discourage creative risk-taking in favor of established conventions, hindering imaginative education. They often perpetuate inequity by ignoring the socioeconomic contexts shaping students. While order and structure have merits, conformity should not come at the expense of nurturing each child's distinct talents. This standardized approach ultimately limits the potential of both students and educators to innovate new models that could profoundly transform early childhood education.

In contrast to the conforming model, the reforming approach involves making incremental changes to policy, curricula, and educational structures to elevate equity and adapt to learners' needs. It is grounded in accountability, flexibility, and personalized learning tailored to each student's strengths and challenges. The reforming approach recognizes that a one-size-fits-all education fails diverse learners.

Education to thrive in an automated world

It aims to improve existing systems by incorporating more individualized instruction, culturally responsive teaching, and data-driven adjustments to better serve students.

This model is gaining popularity for its focus on meeting kids where they are academically and socially rather than expecting them to conform. However, reforming has limitations in only tweaking current conventions versus reimagining learning altogether. Still, its affirmation that access, inclusion, and personalized academic growth matter signals a move in the right direction. The reforming approach also encourages regular critical evaluation of outcomes and programs to enable evidence-based decisions about supportive policies and instruction. While deeper transformation may be needed, the push towards equity and personalized learning makes reforming strategies an important step forward.

Finally, the transforming model represents a fundamental reimagining of education's possibilities, emphasizing radical innovation and sustainability. Rather than mere improvements, transforming seeks to completely rethink schooling from the ground up. This model aims to build learner capacity, creativity, and empowerment through restructuring learning environments to support diverse needs. The focus is on catalyzing deep shifts in educational paradigms towards more equitable, imaginative, student-driven models.

The transforming approach recognizes that thriving in a complex future requires the flexibility to reinvent learning, not just reform existing conventions. Students must be co-creators of their education, not passive recipients of rigid, outdated methods. Transforming schools become centers of discovery, creation, and learner-driven inquiry. They integrate community partnerships and real-world projects tailored to students' passions and purpose.

Assessments evaluate competencies, not just content recall. Teachers facilitate self-directed learning rather than dictate standard curricula. This bold, inclusive approach asks us to dream past the constraints of traditional schooling. While reforming has merits, transforming calls us to revolutionize education to nurture society's creators, healers, leaders, and innovators. It is grounded in faith that placing agency and resources in children's hands can unlock their boundless potential.

As Artificial Intelligence disrupts traditional curricula, striking a balance between reforming and transforming approaches seems prudent. AI can personalize education by analyzing performance and adapting to individual needs. Policies and strategies must be flexible to leverage these emerging technologies. Equity must remain central.

A balanced approach combines the benefits of reforming and transforming models. Maintaining accountability and equity from the reforming paradigm while radically reimagining learning environments as the transforming approach advocates will prepare students for the AI age. This blend can meet diverse learner needs through innovation while providing inclusive, quality education.

Brynjolfsson and McAfee also recomended the promotion of primary education, arguing that education inequality is a "race between education and technology." which suggest that in a world where technology advances too quickly for education to keep up, inequality generally rises. Therefore, it is vital to make substantial investments in primary education.

Acquiring Skills to Thrive in the AI Age

One of my motivations for writing this book was to clearly identify the skills needed to thrive in the era of artificial intelligence, as well as potential courses to develop those capabilities. After extensive research, some key competencies have become apparent.

As AI transforms work, certain technical skills are increasingly essential, including programming, data science, analytics, and design thinking. These skills enable professionals to use intelligent systems efficiently. Hands-on education in computer science, statistics, and related disciplines can equip workers with these capabilities.

However, human skills like creativity, complex communication, and social intelligence remain irreplaceable by machines. Fields relying heavily on human qualities like compassion will maintain this human-centric approach. As AI handles analytical tasks, roles providing a human touch may gain prominence. For instance, as algorithms assume certain clinical responsibilities, we may need more social workers to bring their humanity to community aid.

While technical fluency is mandatory, nurturing humanity's most profound gifts is equally vital. We need holistic education that imparts technical skills alongside teaching imagination, ethics, and abstract thinking. This multifaceted preparation will allow people to complement AI systems and pursue fulfilling roles.

The future is unwritten. By proactively developing diverse technical and human skills, we all can positively shape an automation age where AI augments our potential. With lifelong learning and dedication, workers can thrive amidst the decades of disruption ahead.

The skills defining success in the AI era are continuously evolving. Adaptability and openness to learning are crucial. For those curious and ready to take charge of their destiny in this technology-driven world, immense possibility lies ahead. Let's embark on this journey, equipped with knowledge, questions, and a passion for growth.

I welcome your insights on the skills and mindsets that will propel us all into an era where human imagination remains boundless. Please connect with me at ofeimunfavour@gmail.com as we venture forth together into the promise of the AI age.

CHAPTER SEVEN

THE CAREER OF THE FUTURE

Content Creation and the Socio-Media Revolution

As we venture further into the digital frontier, it becomes increasingly evident that the virtual realm is a world unto itself. Much like the tangible world we navigate daily, the online domain is a bustling hub of activities, with social media standing as one of its most vibrant components. In today's landscape, social media isn't just about connecting; it's a thriving arena where livelihoods are forged through content creation, influence, and digital entrepreneurship. The meteoric rise of social media platforms over the past decade has triggered an insatiable appetite for digital content. Whether it's the addictive allure of TikTok, the visual narratives on Instagram, or the snippets of thoughts on Twitter (now known as X), billions of users are immersing themselves in a virtual world that thrives on content diversity. This digital hunger has given rise to an army of content creators, individuals who weave stories, entertain, educate, and inspire, all while crafting sustainable careers from their passions.

Jimmy Donaldson is one of many who has made a name for himself on social media. The stratospheric growth of the YouTube sensation, popularly known as Mrbeast, exemplifies how social media can enable common individuals to develop profitable professions through creative thinking and demonstrating generosity.

As a teenager, Jimmy began uploading comedic videos and gaming content just for fun, like countless other aspiring YouTubers. He found his passion, despite modest view counts. After years of perseverance, MrBeast had his breakthrough in 2017 through viral stunts like reading the entire dictionary on camera. His outrageous ideas and tireless work ethic allowed him to gain millions of loyal followers. Soon MrBeast leveraged his fame for selfless ends, pioneering epic charity efforts like planting 20 million trees and gifting strangers' houses. This creative generosity cemented his popularity. Though profit was never the goal, MrBeast organically built a flourishing business with branded merchandise and sponsors.

Today MrBeast helms an empire of channels boasting over 100 million subscribers combined. He collaborates with celebs like The Rock while constantly raising the bar on viral entertainment. His enthusiasm inspires fans globally. Still in his early 20s, Jimmy Donaldson exemplifies how through dedication, imagination, and leveraging technology for good, anyone can transform lives. While automation assists with video editing and production, the irrepressible humanity of MrBeast represents the heart of his appeal and impact. He illustrates that content creation remains a profoundly human endeavor - the personalities, relationships, and stories are what ultimately resonate. Although AI enables efficiency, only the human touch can inspire.

Back in the early 2000s, platforms like Facebook and YouTube were merely digital playgrounds where individuals dabbled in content creation for fun. Who could have predicted that this playful endeavor would morph into a booming media industry, providing livelihoods to hundreds of thousands across the globe? With the ascent of content creation, a novel media category has emerged, fundamentally transforming entertainment and giving rise to the lucrative influencer marketing industry, now valued at a staggering $25 billion and still climbing.

This swift expansion of a tech-mediated entertainment realm has not only changed the landscape but has also paved the way for exciting new professional opportunities that were unthinkable just two decades ago. Reflecting on this, I had a discussion with a friend, and I came up with a simple analogy: just as we heard stories of our grandparents through reading or word of mouth, our own tales will be narrated to our grandchildren through the content we create. In this digital age, storytelling has taken on a vibrant form, driven by reels, videos, and posts.

For entrepreneurs, content creation has evolved into a fundamental testament of their work. Your social media presence has become a modern-day portfolio, showcasing your credibility and offering a glimpse into your professional world. Potential clients and partners are likely to scrutinize your online presence, drawn by the stories you share and the value you provide. Even within the realm of traditional employment, the significance of social media cannot be overstated. Employers now turn to these platforms as windows into the lives and personalities of prospective employees. It's a contemporary form of assessment, one that goes beyond the confines of a resume. Your online identity and content serve as a virtual introduction, often influencing decisions in the hiring process.

All this highlights the immense value that the social media landscape holds today. It's a realm where creative prowess can carve out extraordinary paths. Individuals can etch their stories onto the walls of fame or choose their own unique narrative. The power to create, influence, and leave a digital legacy has become a defining characteristic of our generation. As the curtains of the digital age continue to rise, the stage is set for storytellers and creators to shine.

For creative individuals who resonate with specific social platforms, this represents a monumental opportunity. Whether drawn to short-form videos, livestreaming gameplay or producing episodic content, budding creators can leverage these apps to find and engage niche audiences. If they consistently provide value through imagination and authenticity, the potential to build an audience of loyal followers is immense. While discovering the winning formula takes commitment and perseverance, the lives of internet celebrities attest that building a profitable full-time career solely around social content is attainable for those with drive and determination. This new media landscape promises to reward creativity, digital fluency and human connection. The next generation of rising stars will be internet trailblazers who can entertain and inspire online communities.

The Untapped Potential of Education in the Digital Age

In the dynamic tapestry of education, there exists a realm that is often sidelined, labeled as a mere distraction - the vast and intricate world of social media. It's a domain where screens replace chalkboards, where connections span continents, and where influence is measured in likes and shares. Yet, despite its ubiquity, the formal education system has largely turned its back on this realm, missing a golden opportunity to empower the next generation for the future.

Historically, the educational landscape has viewed social media through a lens of skepticism. It's often seen as a perilous distraction, diverting young minds from the traditional pursuits of academia. Social media, in this context, is relegated to the role of a villain, a force that tugs at attention spans and undermines scholastic achievements. However, there's an alternate perspective waiting to be explored.

Imagine a scenario where social media isn't just a realm of selfies and hashtags, but a realm of profound learning, connectivity, and self-expression. Instead of stifling its influence, formal education institutions can harness this digital realm as a powerful tool for positive impact. Just as children are taught the intricacies of mathematics and the nuances of science, they can also be guided to navigate the digital world wisely.

The case for integrating social media education into the curriculum is strong. The world is evolving, and the future of a nation is intrinsically tied to the preparedness of its youth. Teaching them to wield social media effectively, to share their unique abilities, and to discern credible sources from misleading ones is not just a luxury; it's a necessity. In an era where digital literacy is as vital as traditional literacy, empowering children with these skills is akin to equipping them with armor for the digital battlefield.

Yet, it's crucial to acknowledge the challenges that tread alongside the benefits. The social space, like any domain, has its shadows. In an environment where highlight reels dominate, feelings of inferiority can sprout like weeds in young minds. The constant comparison between one's journey and the perceived success of others can breed a stalk of depression, overshadowing the joys of real-life achievements. The pressures to conform, to change deeply-rooted philosophies and ideologies to appease an online audience, are palpable. And then there's the menace of cyberbullying, a dark corner of the digital world where cruelty thrives under the cover of anonymity.

Amidst these pitfalls, the remedy lies not in shunning the digital realm, but in cultivating self-awareness and gratitude. Just as the heart of a storm is calm, self-awareness is the anchor that holds steady amidst the tumultuous waves of comparison. It's understanding that worth isn't measured in likes or retweets, but in the genuine impact we make on others and the world around us.

Gratitude, on the other hand, is the antidote to the poison of envy. Teaching ourselves and our children to embrace gratitude is a transformative step. It reminds us that every accomplishment, no matter how small, is a victory worth celebrating.

In the age of artificial intelligence, where machines learn faster and process more data than humanly possible, there's an eternal truth that remains unshaken: the essence of being human. Amidst the algorithms, the data streams, and the machine-led transformations, there's a space that is inherently ours - the realm of self-awareness, creativity, empathy, and gratitude. Here, we must teach our children to find their footing, to navigate the digital world with wisdom, and to embrace their uniqueness amidst a sea of sameness.

In this narrative, formal education isn't just about algebra and history; it's about imparting life skills that navigate the digital age. It's about recognizing that social media isn't an adversary, but a canvas for expression, a bridge for connection, and a powerful platform for positive change. It's about empowering our children to use the digital tools at their disposal not just for selfies, but for building meaningful connections, for fostering empathy, and for shaping a future where humanity thrives, even amidst the rapid advancements of AI.

As algorithms continue to evolve and automation reshapes various sectors, one truth remains unwavering - the future of socio-media, of content creation, and of our collective journey in this digital era will be determined not just by technology, but by the power of the human spirit. It's a journey where the stories we share, the connections we foster, and the impact we make will echo across the digital expanse and shape the landscape for generations to come.

Finding Success on Creator Platforms

Today's digital landscape offers an array of platforms for content creators to build an audience and potentially profit. But simply trying to make money right away is not the best approach on these platforms. First, you must lay the groundwork for success by focusing on creating quality content consistently, developing a unique brand, and forming meaningful connections with followers. The most successful creators on sites like YouTube, Instagram, and TikTok got their start by posting content that brings value to viewers. They carved out a specific niche and shared their passions authentically. Building a loyal audience comes first. Monetization follows later as a natural extension of that engaged community.

Once a creator has cultivated a sizable, devoted following, there are plenty of built-in tools on each platform to unlock monetary opportunities. A clear testament to the potential of this monetization is embodied in the likes of global celebrities such as Cristiano Ronaldo. For many, he's the celestial body of football, dazzling fans with his skills. On platforms like Instagram, where he commands a following of nearly 600 million (as per the last count), he becomes a prime candidate for brand endorsements. Ronaldo's stardom, amplified by his digital presence, enables him to command an average post price of a staggering $2,397,000. But beyond the luminaries like Ronaldo, countless stars have harnessed the power of their audience, partnering with brands that align with their niche.

However, endorsements are just one aspect of the monetization universe. Direct fan funding has emerged as a revolutionary force, offering creators a revenue stream that extends beyond traditional advertisements. Platforms now enable fans to financially appreciate creators directly. Features like Super Chats on YouTube allow fans to make their comments more visible, for a price. Super Stickers serve as virtual tokens of appreciation during live streams, while initiatives like Super Thanks act as direct tipping mechanisms.

Channel memberships further exemplify the potential of fan-based revenue. Followers, driven by their admiration and loyalty, willingly subscribe to membership tiers, unlocking exclusive content, livestreams, merchandise discounts, and other enticing perks. To put things in perspective, MrBeast, a content creator, reportedly rakes in $250,000 a month purely from these memberships.

Last but not least, affiliate marketing remains an evergreen avenue for creators. By embedding relevant product links within video descriptions or content, creators can earn commissions. Many have evolved this approach, setting up comprehensive affiliate stores to enhance potential earnings. However, the key is ensuring this commerce flows organically from the relationship creators have built with their community. It should never feel forced or take away from the quality content that attracted fans in the first place.

Success on today's creator platforms requires patience, consistency and valuing your audience above all else. While earning money from one's passion is an amazing opportunity, it is not guaranteed overnight. But for those ready to put in the work, share their gifts, and cultivate authentic human connections, the digital world offers endless possibilities.

Platforms Driving the Ecosystem

In the modern digital era, platforms such as YouTube, TikTok, and the reinvented Twitter, now known as X, have emerged as not just social hubs but as powerful economic drivers for content creators. These platforms have democratized content creation, offering tools and monetization opportunities that can translate creativity into substantial revenue.

YouTube, for instance, has always been a forerunner in this revolution. Once creators gather momentum by reaching the milestone of 1,000 subscribers and garnering 4,000 watch hours within a year, they can tap into the YouTube Partner Program. This isn't just a badge of honor; it opens the doors to monetization features that can significantly bolster one's income.

The pivotal role of advertisements in this paradigm cannot be emphasized enough. With ads being showcased before, during, and after videos, creators are poised to earn impressive amounts. Especially when they leverage a plethora of ad formats such as display, overlay, and bumper ads. It's not unheard of for top creators to amass earnings of over $5 for every 1,000 views. A simple extrapolation suggests that if a creator can enchant their audience enough to get a billion views, a staggering $5 million awaits them.

Similarly, TikTok, which on first glance seems predominantly about catchy dance routines and concise comedic sketches, is a goldmine for those who know how to navigate its monetization maze. At the very beginning, once a creator crosses the threshold of 1,000 followers, they're introduced to the TikTok Creator Fund. This fund, amounting to $200 million, ensures that creators are compensated based on their content's performance, roughly around 2-4 cents for every 1,000 views. As creators ascend in popularity, amassing 10,000 followers and beyond, the rewards become richer.

TikTok LIVE gifting emerges as a lucrative channel, where ardent followers can offer virtual gifts that vary in monetary value. On top of this, brand collaborations become feasible, providing a steady flow of income. When creators breach the 100,000 followers mark, they're not just popular, they're influencers. With this coveted status comes an array of exclusive monetization opportunities that only become more lucrative as one's follower count swells, especially when it reaches the realms of 500,000 or even a million.

Yet, the landscape of content monetization is not just dominated by video platforms. X, evolving from what was once Twitter, presents an innovative avenue for its users to derive income. This initiative is especially tailored for its committed users, particularly those who've embraced its premium offering, formerly known as Twitter Blue.

To ensure that quality is at the heart of this initiative, X has put in place stringent criteria. It's not just about having a formidable presence, illustrated by metrics such as 15 million organic impressions over a quarter or a robust follower count exceeding 500. The platform is intent on curating creators who uphold its values, epitomized by authenticity, security, and alignment with X's policies. After aligning with these prerequisites, the journey to monetization on X is seamless. Once creators integrate with a Stripe account, the platform chosen by X for financial transactions, they're on their way to earn from their content.

In wrapping up this chapter, it's crucial to recognize a transformative shift in our understanding of future careers. Content creation, previously seen as a hobby or sideline, is rapidly taking center stage as the profession of the future. As the lines between digital communities and financial realms intertwine more closely than ever, platforms like YouTube, TikTok, and X are not just recreational spaces; they are vibrant economic arenas. These platforms provide real, accessible avenues for individuals to morph their creativity and zeal into rewarding professions. And while artificial intelligence can assist and amplify human endeavors, it can never replace the authentic passion and innovation that is inherently human. As we look forward, the future of content creation seems boundless, driven by the individual's spirit and ingenuity.

RECOMMENDATIONS AND ETHICAL CONSIDERATIONS

Policy's Role in Guiding an Automated Workforce Transition

In previous chapters, I stressed the necessity of individuals recognising their unique abilities, tapping into their inner creativity and sensitivity, and building meaningful connections. These personal characteristics are critical for people to thrive in the automation era. However, in order to guide this transitional moment with accountability and foresight, the government must support a cohesive strategy, encouraging collaboration among itself, industry, and academic institutions.

The changes ahead require policy foresight and coordination from all societal stakeholders. We must move beyond partisan bickering and siloed interests toward open forums where diverse viewpoints and expertise jointly inform wise governance of emerging technologies. Neither side of the political aisle has a monopoly on solutions. Progress requires bringing together policymakers, technologists, ethicists, futurists, workers and more to debate challenges, propose thoughtful reforms, and ensure AI develops in the service of shared human values.

In his insightful book "Rise of the Robots," Martin Ford emphasizes the critical importance of collaboration between key stakeholders to responsibly steer automation's impacts. Ford argues governments must actively coordinate with industry and education groups to align training programs with real-world workforce needs. As industries drive much of the innovation, they should collaborate with academia to equip students with in-demand skills for applying emerging technologies. Ongoing public-private partnerships can help education stay responsive to industry trends, ensuring graduates have opportunities to utilize their knowledge.

Additionally, policymakers, private companies, and researchers should collaborate to promote ethical and socially-conscious technological development. By working together on forward-looking policies and innovation ecosystems, stakeholders can maximize benefits of new technologies while minimizing harms to workers and communities. Supporting R&D initiatives and startups focused on human-centric automation can catalyze job creation. Testing new applications in controlled environments allows potential risks to be addressed proactively before widespread deployment. No single group can navigate the transition alone. Governments can provide guidance and support, industry offers technical insights, and academia supplies critical evaluation and skills training.

By recognizing their shared responsibility, these factions can constructively shape automation for the common good. Open and ongoing multilateral collaboration will drive both ingenuity and inclusivity as existing jobs evolve and new ones emerge. Working as partners, not isolated entities, is imperative to realize the best possibilities of the automation age.

Equity not equality

Collaboration should prioritize equity over equality. Equity emphasizes recognizing and catering to the unique needs and challenges of diverse groups. This perspective is especially crucial when addressing the impact of automation on the workforce. As we have seen, automation often affects different individuals and communities differentially. While some people may be able to adapt to new technologies and find employment opportunities, others may struggle to keep up or to find new jobs that require different skills. By focusing on equity, policymakers can ensure that all groups have access to the opportunities and resources they need to thrive in an age of automation, rather than simply providing equal outcomes across the board. This means considering how automation will impact different populations, including communities of color, low-income workers, and the elderly.

A significant proposition put forth by experts like Frank Levy and Richard J. Murnane is the introduction of compensation initiatives for those adversely affected by automation's ripple effects. Essentially, these initiatives aim to cushion the blow for individuals facing the brunt of rapid technological advancements.

Imagine a scenario where automation displaces a factory worker or a cashier. Instead of leaving them to grapple with unemployment, compensation programs could step in, offering a range of support. This could be in the form of direct financial assistance, opportunities for retraining, or even educational resources to help them pivot to new roles.

The overarching objective is clear: equip individuals with the tools and resources they need to navigate and prosper in this new economic terrain.
But how do we fund such ambitious programs? One avenue could be through governmental channels, possibly by levying taxes on automation-centric industries. A concept gaining traction in some circles is the idea of an "automation tax" that could potentially fund a universal basic income, ensuring a safety net for those affected. On the other hand, the private sector, particularly affluent tech magnates, has shown interest in philanthropic endeavors. Contributions from these individuals and entities, especially those who've profited immensely from automation, could support sectors most at risk, like retail and manufacturing.

In the past few years, we have seen an increasing number of philanthropic pledges made by leaders in the technology industry. These generous donations are aimed at tackling major societal issues like education, economic inequality, and poverty.

One major gift was announced in 2017 by Amazon's CEO Jeff Bezos. He committed $1 billion to establish a network of tuition-free preschools in underserved communities across the United States. Bezos hoped this initiative would give children from low-income families access to high-quality early education, which research shows is crucial for future academic success.

In 2019, Salesforce CEO Marc Benioff pledged to donate $1 million annually to economic justice causes until his retirement. The beneficiaries of this long-term commitment are civil rights organizations and other groups working to close the economic divide. Benioff has been outspoken about the need for corporations to help address income inequality.

Also in 2019, Microsoft co-founder Bill Gates earmarked a massive $158 million donation through his philanthropic Bill & Melinda Gates Foundation. The funds were specifically allotted to initiatives aimed at reducing poverty and improving socioeconomic mobility in the United States. This built on the foundation's extensive previous work to lift disadvantaged populations out of poverty globally. These recent major pledges by Bezos, Benioff and Gates demonstrate the vast sums tech leaders are willing to commit to solving systemic issues of inequality, lack of access to education, and poverty. Their long-term donations seek to make lasting change. It sets an example for the industry to use its wealth and influence for the public good.

These instances are just the tip of the iceberg. Many other tech magnates and philanthropists have made significant contributions to causes impacted by automation, even if they haven't always been in the limelight. While relying solely on philanthropy isn't a silver bullet, it could be a significant piece of a multifaceted solution, complemented by government policies, private sector investments, and robust educational initiatives.

Let's delve into some numbers. According to a report from the McKinsey Global Institute, up to 800 million global workers could be replaced by robots by 2030. Such staggering statistics underscore the urgency of implementing compensation programs. If managed effectively, these programs could be a game-changer in bridging economic disparities.

The beauty of compensation initiatives lies in their potential to democratize the benefits of automation. By ensuring that individuals have the resources and opportunities to adapt, we're not just providing a safety net; we're fostering an environment where everyone has a fighting chance to succeed in the face of relentless technological progress. In essence, these programs could be the linchpin in creating a more inclusive and equitable future, where the promise of automation is realized by all, not just a select few.

The Role of Strategic Fiscal Policies

One approach to mitigate some of these negative impacts is through strategic fiscal policies, which could encourage growth and development to make sure the benefits of technological progress are broadly shared. Rooted in fiscal policy, emerges as a beacon of hope wich is the strategic use of tax incentives. Tax incentives could be an impactful policy lever to ensure automation's benefits are shared inclusively across society.

For instance, governments could offer tax credits to companies that invest in retraining and hiring workers displaced by technological changes. This motivates businesses to prioritize workforce development alongside efficiency gains from automation. Studies show such incentives are effective. A recent Cornell study found a $1,000 tax credit increased employer-provided training among eligible workers by 10%. Germany's Employment Agency provides wage subsidies up to 50% for 6 months to employers who hire and train unemployed individuals. This was linked to long-term employment gains even after the subsidy ended.

Meanwhile, leaders in adopting automation could be subject to a surtax. Globe Telecom, a leading Philippines telecom pioneering automation and AI, has funded employee retraining programs through a 1% increase in corporate income tax since 2019. The revenue specifically sponsors scholarships for digital skills. Similarly, Thailand imposed a 0.001% surtax on e-commerce transactions to support digital training programs.

This progressive tax structure shares the burden equitably. Those benefiting most from automation fund programs to uplift disadvantaged groups. It also disincentivizes overeager automation by making firms accountable for societal impacts. Companies become motivated to implement automation responsibly and reskill displaced workers.

Workers too should receive tax help for self-directed learning. Deductions for job-related education exist but could be expanded through tax credits or lifelong learning accounts, which provide an annual use-it-or-lose-it fund that workers draw from to pay for training. Singapore's SkillsFuture program exemplifies this model.

Overall, incentives and surtaxes allow governments to calibrate automation's pace while protecting workers. Leaders enjoy profits but also invest in common good. Workers receive support to transition into new roles. Collective responsibility coupled with enlightened self-interest creates an automation pathway that leaves no one behind. Fiscal policy becomes a lever to constructively shape technology's trajectory.

Entrepreneurship in the Age of Automation

Another vital area for policy intervention is spurring entrepreneurship and small business creation. Entrepreneurs are naturally inclined to think creatively and find novel ways to solve problems.

Additionally, successful entrepreneurs don't just create new businesses - they also create new opportunities for workers, new industries, and new markets. For example, Thomas Edison and Henry Ford aren't just remembered for creating successful businesses, they are also remembered for creating entire industries and jobs that didn't exist before.

As automation transforms existing jobs and industries, entrepreneurship will become an increasingly significant avenue for displaced workers to create new economic opportunities. Self-employment and small business funding should be encouraged as productive outlets for workers navigating career transitions.

Governments have a powerful role in shaping the entrepreneurial landscape. By implementing the right policies, they can ignite the entrepreneurial spirit and provide the tools needed for success. One way they can do this is by funding programs like startup accelerators and incubators. These programs offer budding entrepreneurs the guidance and resources they need to turn their ideas into successful businesses.

Additionally, vocational programs that focus on business skills can equip individuals with the knowledge they need to navigate the business world.

Tax incentives can also make a significant difference. By allowing startups to defer certain costs and offering credits for research and development, governments can ease the financial burden on new businesses. Moreover, providing grants and loans with favorable terms can offer startups the capital they need to get off the ground. Programs like the SBA's Small Business Innovation Research initiative are excellent examples of how funding can be directed to startups that are working on cutting-edge technologies.

Looking at Silicon Valley, we see the impact of such policies. Between 2005 and 2015, over half of the job growth in the region came from new startups and young firms. This growth was facilitated by policies that made it easier for entrepreneurs to start and grow their businesses. Research, like the Global Startup Ecosystem Report in 2020, has shown that cities with good funding access, mentorship opportunities, and a strong startup culture tend to see more growth in ventures and job creation.

Singapore offers another shining example. The country has made significant strides in becoming a global startup hub by focusing on education and providing grants and incubation schemes. Their recent allocation of over $1 billion to help small businesses move their operations online shows their commitment to supporting businesses. While it's essential to be cautious and select ventures with high potential, such support can lead to significant business and job growth.

However, in this age of automation, adaptability is crucial for entrepreneurs. As Erik Brynjolfsson and Andrew McAfee point out, as more tasks become automated, entrepreneurs need to be flexible. They should be prepared to pivot from industries heavily affected by automation and look for opportunities where technology enhances human capabilities. This might mean adapting their skills or even learning new ones to fit into changing markets. But it's also essential to note that not all entrepreneurial efforts are the same. While technology offers vast opportunities, it can also lead to markets dominated by a few big players. Brynjolfsson and McAfee caution against entrepreneurs who might not add value or even engage in questionable activities. It's vital for entrepreneurs to focus on creating value for society as a whole.

Ethics, Responsibilities, and the Human-AI Partnership

In the rapidly evolving world of Artificial Intelligence, business leaders and executives must tread with care. While these tools offer unprecedented capabilities, it's essential to continuously evaluate and ensure that the generated outcomes meet our standards of quality and satisfaction. Just because an AI, like ChatGPT, can produce content, doesn't automatically qualify it as exceptional. Similarly, generating a visual landscape using AI doesn't necessarily mean it's fit for the final cut of a film.

As we navigate this transitional phase, co-creating with algorithms, it's imperative for leaders, especially those at the helm of AI-driven companies, to refine their decision-making skills. A foundational question should always be: "Who benefits from our AI tools?" Our guiding principles should be rooted in transparency, fairness, empathy, and responsibility.

To institutionalize these values, companies should consider establishing an ethics board or council. This body would serve as the moral compass, ensuring the ethical integration of generative AI.

Furthermore, it's vital to equip all team members with ethical training on AI usage. This education should address not only the effective use of AI but also help individuals navigate their apprehensions, challenges, and potential biases towards this advanced tool.

As technology progresses, the lines distinguishing human-generated content from AI-produced content will likely blur. This makes it even more crucial for leaders to discern the roles and contributions of each. The key is to ensure that human consciousness remains central to AI endeavors. By doing so, we can guarantee that AI-generated content aligns with organizational values and objectives, ultimately serving humanity's betterment.

Engaging deeply with generative AI, understanding its strengths and limitations, ensures we don't relinquish decision-making power to machines. The aim is a harmonious blend: harnessing AI's capabilities to amplify human creativity while retaining human oversight.

A prevalent misconception is that AI's most significant bias revolves around race, ethnicity, or gender. However, the real bias is humanity's self-perceived inferiority. Elevating machines while undermining human capability gives AI undue authority. While phrases like "AI created this artwork" or "AI is evolving rapidly" are popular, it's vital to remember that humans are the architects behind AI. We conceptualize, curate, and supervise AI to achieve specific results. If we position AI as the core of our processes, we risk sidelining human contributions, potentially paving the way for a future where human roles become redundant.

Instead, our narrative should emphasize humans' central role in AI's creation and application. While it's true that AI tools can produce art or advance technologies, it's humans who harness these tools for creative expression and innovation. By designing AI tools, we inadvertently transfer our judgments, insecurities, and limitations onto them. Thus, overcoming our self-doubts and viewing AI as an augmentative tool, rather than a competitor, is crucial. This perspective ensures that AI systems uplift humanity, enhance our creative capacities, and help us realize our collective potential.

CONCLUSION

Embracing AI with Humanity at its Core

As we wrap up this enlightening journey through the realm of AI, the question that naturally arises is, "What's next?" The path forward is clear: it's time to confront and conquer the apprehensions, judgments, and misconceptions surrounding AI. I deeply resonate with the concerns many express about AI. The looming anxieties of AI taking over jobs, replacing human roles, and redefining our identity in the workplace are sentiments I encounter frequently. However, it's essential to ground ourselves in the fundamental truth: AI is merely a tool designed to serve humanity, to serve you.

To dispel fears, we must expand our horizons. How do we achieve this? Firstly, by immersing ourselves in continuous learning. Beyond this book, delve deeper into the symbiotic relationship between AI and human potential. Reflect on the unique qualities that set humans apart from machines and strategize on amplifying these attributes.

Next, it's time to spring into action. As underscored in this book, the essence lies in recognizing our strengths, understanding our unique capabilities, and aligning them with the AI tools at our disposal. My experience penning this book stands as a testament to this principle. Despite never having authored a book or mastered professional English writing, I've successfully crafted this work, thanks to a plethora of AI tools that aided in research, editing, and auto correction. The secret? Staying engaged, adaptive, and forward-thinking.

Creativity will be our guiding star. One of the imminent challenges we face is the risk of settling into a realm of mediocrity. While AI is a remarkable tool, it operates within set parameters, much like a camera that captures images in a consistent manner. If we lean too heavily on AI, neglecting our innate creative flair, the results, though efficient, might lack the unique touch of human brilliance.

The world of AI is dynamic, evolving at a pace that's almost dizzying. As John Finger, an AI enthusiast, aptly puts it, we're navigating an era of "cutting-edge obsolescence." Today's groundbreaking innovation might be tomorrow's outdated tech. To stay ahead, it's crucial to surround ourselves with a reservoir of resources, ensuring we're always in tune with the latest developments.

In closing, I extend my heartfelt gratitude for accompanying me on this enlightening expedition. Remember, in the dance between humans and AI, it's the human spirit that leads. You are, and always will be, the driving force behind the marvels of AI.

NOTES

Chapter 1

Ford, M., 2015. Rise of the Robots: Technology and the Threat of a Jobless Future. Basic Books.

Brynjolfsson, E. and McAfee, A., 2014. The second machine age: Work, progress, and prosperity in a time of brilliant technologies. WW Norton & Company.

Bardon, A., 2013. A brief history of the philosophy of time. Oxford University Press.

McKeown, G., 2014. Essentialism: The disciplined pursuit of less. Currency.

MITS Altair 8800: Wikipedia page on Altair 8800

Apple II: Wikipedia page on Apple II

IBM PC: Wikipedia page on IBM Personal Computer

MS-DOS: Wikipedia page on MS-DOS

Apple's Macintosh: Wikipedia page on Macintosh

World Wide Web: Wikipedia page on World Wide Web

Netscape Navigator and Microsoft's Internet Explorer: Wikipedia page on Netscape Navigator

Wikipedia page on Internet Explorer

Chapter 2

Google Analytics (analytic tool for websites) Official Page

Apple Watch (reference to health features and data collection) https://nr.apple.com/DH3i2T3Eb6

IBM's Deep Blue (chess-playing computer) https://www.ibm.com/ibm/history/ibm100/us/en/icons/deepblue/

Siri and Alexa (digital personal assistants) https://www.forbes.com/sites/johnkoetsier/2020/08/08/alexa-siri-google-assistant-how-the-top-smart-assistants-stack-up/

Google's search engine (uses machine learning algorithms) https://www.searchenginejournal.com/ml-things-we-know/408882/#close

Chapter 3

Oxford University researchers - Frey, C.B. and Osborne, M.A., 2017. The future of employment: How susceptible are jobs to computerisation?. Technological forecasting and social change, 114, pp.254-280.
Automated Teller Machines (ATMs) - Referenced in the context of the financial industry's transformation.
https://www.ncr.com/blogs/banking/history-atm-innovation
Foxconn - Mentioned as a significant Apple supplier that has deployed robots in their plants. https://www.idownloadblog.com/2016/05/25/foxconn-robots-success/
ABB - Recognized as a pioneer in the field of AI, with a mention of their ABB Ability solution. https://www.abb-conversations.com/2022/12/getting-our-house-in-order-with-ai/
General Electric, Kuka, and Siemens - Manufacturing giants that have embraced AI for various tasks. https://www.oreilly.com/content/adopting-ai-in-the-enterprise-general-electric/
https://aiexpert.network/case-study-how-siemens-is-transforming-supply-chain-with-ai/
https://www.kuka.com/en-de/company/iimagazine/2022/artificial-intelligence-in-production
Tesla - Mentioned in the context of using over 1,000 robots in its factory.
https://www.roboticstomorrow.com/article/2022/06/2022-top-article-how-tesla-used-robotics-to-survive-production-hell-and-became-the-worlds-most-advanced-car-manufacturer/18908#:~:text=They%20purchased%20over%201%2C000%20robots,tasks%20like%20wire%20harness%20assembly.
Delta Airlines - Referenced for its use of facial recognition software in airports. https://www.forbes.com/sites/jenniferleighparker/2021/10/27/first-look-delta-tsa-launch-facial-recognition-at-atlanta-airport/
EasyJet - Highlighted for its use of data science in reducing food waste and driving profitability. https://sustainabilitymag.com/esg/easyjet-leverages-sustainability-ai-to-reduce-food-waste
SESAR JU - Mentioned in the context of funding projects for AI-powered tools in air traffic management. Frey, C.B. and Osborne, M.A., 2017. The future of employment: How susceptible are jobs to computerisation?. Technological forecasting and social change, 114, pp.254-280.
AI in healthcare https://www.youtube.com/watch?v=G1IsZeFR_Rk&t=59s
Dr. Bertalan Mesko - A renowned Medical Futurist, cited for his video titled "How COVID-19 Changed The Future of Healthcare.
https://www.youtube.com/watch?v=3uWSuK8Wu0o&t=523s
Sensely - The company behind the virtual health assistant named Molly.
https://www.youtube.com/watch?v=AU1nGpOmZpQ
Resemble AI - Software platform for voice cloning.
https://www.resemble.ai/cloned/

Drake and The Weeknd - Mentioned in the context of a voice-generated song on TikTok. https://www.resemble.ai/ai-voice-rap/

Amper and AIVA - Companies that have developed AI tools for music composition. https://www.aiva.ai/

Google Labs' Magenta - Offers free plugins to Ableton users. https://blog.bpmmusic.io/news/the-future-of-ai-and-music-production/

Izotope's Neutron and Ozone suites - Powered by AI for mixing and mastering. . https://blog.bpmmusic.io/news/the-future-of-ai-and-music-production/

LANDR - Used for machine learning-based mastering. https://blog.landr.com/ai-in-music/#:~:text=LANDR%20Mastering&text=Our%20Synapse%20mastering%20engine%20uses,audio%20features%20of%20your%20track.&text=Based%20on%20what%20it%20learns,and%20refined%20by%20human%20engineers.

Adobe's Sensei - Leveraged for automating complex tasks in design. https://news.adobe.com/news/news-details/2016/Adobe-Sensei-Lets-Customers-Master-the-Art-of-Digital-Experiences/default.aspx

Quizlet - A study platform that uses AI for recommendations. https://quizlet.com/features/learn

Carnegie Learning's Mika software - Adapts to individual student needs. https://www.carnegielearning.com/solutions/math/

Kritik - Uses AI to provide feedback on student performance. https://www.kritik.io/

Pearson - Educational publisher utilizing AI for student performance data analysis. https://plc.pearson.com/en-GB/news/pearson-launches-ai-summer-reading-list

Chapter 4

The trolley problem, created in 1967. https://www.youtube.com/watch?v=Oh760agkTmI&t=5s

Kurzgesagt episode titled "What Is Intelligence? Where Does It Begin?" https://youtu.be/ck4RGeoHFko

Robinson, K. and Aronica, L., 2009. The element: How finding your passion changes everything. Penguin.

Genesis 1:1, 1:27, 1:31, and 2:2 from the biblical account of creation.

Chapter 5

Grant, A., 2013. Give and take: A revolutionary approach to success. Penguin."Leaders Eat Last" by Simon Sinek

Goleman, D., 2020. Emotional intelligence. Bloomsbury Publishing.

Clear, J., 2018. Atomic Habits: the life-changing million-copy# 1 bestseller. Random House.

William Shakespearehttps://www.gabekwakyi.com/essays/what-william-shakespeare-means-by-there-is-nothing-either-good-or-bad-but-thinking-makes-it-so
https://hbr.org/2022/11/how-generative-ai-is-changing-creative-work
https://assets.publishing.service.gov.uk/government/uploads/system/uploads/attachment_data/file/1023590/impact-of-ai-on-jobs.pdf
Healthcare (Context: Importance of human touch and empathy).
https://www.kevinmd.com/2023/02/why-human-touch-matters-in-health-care-the-limitations-of-ai.html

Chapter 6
Sydney J. Harris https://medium.com/the-black-coffee/the-whole-purpose-of-education-is-to-turn-mirrors-into-windows-5b115368f1cd
Udemy, Alison, Coursera, and Google Skillshop:
https://zapier.com/blog/online-professional-development/
https://grow.google/intl/en_in/learn-skills/
Brynjolfsson, E. and McAfee, A., 2014. The second machine age: Work, progress, and prosperity in a time of brilliant technologies. WW Norton & Company.
Clear, J., 2018. Atomic Habits: the life-changing million-copy# 1 bestseller. Random House.
Levy, F. and Murnane, R.J., 2004. The new division of labor: How computers are creating the next job market. Princeton University Press.
Mac Naughton, G., 2003. Shaping early childhood: Learners, curriculum and contexts. McGraw-Hill Education (UK).Neville Goddard

Chapter 7
TikTok, Instagram, Twitter (now known as X): Social media platforms that have become hubs for content creation and digital entrepreneurship.
https://blog.hubspot.com/marketing/new-social-media#:~:text=Platforms%20like%20TikTok%20and%20Instagram,different%20channels%20to%20diversify%20traffic.
Jimmy Donaldson (MrBeast): https://www.bbc.co.uk/news/technology-66372679
The Rock and MrBeast collaborates: https://youtu.be/S_CUEOBZ0P4
Cristiano Ronaldo https://www.goal.com/en-gb/news/cristiano-ronaldo-first-ever-instagram-landmark-lionel-messi-selena-gomez-kylie-jenner/blte341cf4226dd247e
YouTube Monitization: https://tcrn.ch/3kjvgcl
TikTok Creator Fund: https://newsroom.tiktok.com/en-gb/tiktok-creator-fund-your-questions-answered
X (formally twitter): https://techcrunch.com/2023/08/11/elon-musk-twitter-everything-you-need-to-know/

Ford, M., 2015. Rise of the Robots: Technology and the Threat of a Jobless Future. Basic Books.

Marcelin, I., Brink, D., Fadiran, D., & Amusa, H. (2019). Subsidized labour and firms: investment, profitability, and leverage.. https://doi.org/10.35188/unu-wider/2019/684-5

Mohnen, P. (2010). What does it take for an r&d tax incentive policy to be effective?., 33-58. https://doi.org/10.7551/mitpress/9780262014687.003.0002

Stephan, G. Employer wage subsidies and wages in Germany: empirical evidence from individual data. ZAF 43, 53–71 (2010). https://doi.org/10.1007/s12651-010-0029-3

Dominguez-Péry, C. and Vuddaraju, L. (2020). From human automation interactions to social human autonomy machine teaming in maritime transportation., 45-56. https://doi.org/10.1007/978-3-030-64861-9_5

Kemp, J. (2020). Empirical estimates of fiscal multipliers for south africa.. https://doi.org/10.35188/unu-wider/2020/848-1

Ochmann, J. and Laumer, S. (2020). Ai recruitment: explaining job seekers' acceptance of automation in human resource management., 1633-1648. https://doi.org/10.30844/wi_2020_q1-ochmann

Willis, M. and Jarrahi, M. (2019). Automating documentation: a critical perspective into the role of artificial intelligence in clinical documentation., 200-209. https://doi.org/10.1007/978-3-030-15742-5_19

Manyika, J., Lund, S., Chui, M., Bughin, J., Woetzel, J., Batra, P., Ko, R. and Sanghvi, S., 2017. Jobs lost, jobs gained: What the future of work will mean for jobs, skills, and wages.

Genome, S., 2020. The global startup ecosystem report GSER 2020. Startup Genome, San Francisco..

Brynjolfsson, E. and McAfee, A., 2014. The second machine age: Work, progress, and prosperity in a time of brilliant technologies. WW Norton & Company.

INDEX

ABB
Ableton
Abraham Lincoln
Acceleration
Account Problem
Accountability
Accuracy
Acellular Slime Mold
Achievement
Acquiring
Actions
Adapt
Adaptability
Adobe's Sensei
Adolf Hitler
Adoption
Advanced Sensors
Advancements
Aesthetic Judgment
After
Age
AI
AI Education
AI-Powered World
Air Travel
Airline Industry
Airplane
AIVA
Albert Einstein
Alexa
Alexis De Tocqueville
Alignment
Alison
Alongside
Also
Altair 8800
Alternatively
Always
Amazon
Amidst

ABB
Ableton
Abraham Lincoln
Acceleration
Account Problem
Accountability
Accuracy
Acellular Slime Mold
Achievement
Acquiring
Actions
Adapt
Adaptability
Adobe's Sensei
Adolf Hitler
Adoption
Advanced Sensors
Advancements
Aesthetic Judgment
After
Age
AI
AI Education
AI-Powered World
Air Travel
Airline Industry
Airplane
AIVA
Albert Einstein
Alexa
Alexis De Tocqueville
Alignment
Alison
Alongside
Also
Altair 8800
Alternatively
Always
Amazon
Amidst
Amper
Amplify
An
Analogy

Analytical
Analytics
Anchor
Andrew
Antidote
Apollo 13 Astronauts
Apple
Apple II
Apple Watch
Approach
Approaches
Apps
Aptitude
Architects
Arena
Armor
Army
Art
Artificial
Artificial Intelligence
Artificial Intelligence In Creative Arts
Artificial Intelligence In Education
Artificial Intelligence In Health Care
Artificial Intelligence In Web Development
Artisans
Artists
Arts
Assessment
Assignments
Assistance
Assistant Teacher
Astronauts
At
Atms
Atomic Habits
Attributes
Audiences
Augment
Augmentative
Augmented Reality
Augments
Authenticity
Author
Authority

Authors
Automation
Automation In The Aviation Field
Automation In The Transportation Industry
Autonomous
Available
Avenues
Backgrounds
Balances
Barriers
Bases
Batman
Battlefield
Become
Bedside Manner
Benefits
Bertalan Mesko
Best
Beyond
Bias
Biases
Big
Bill
Billion
Billion Views
Binary
Biology
Birthplace
Blacksmithing
Blend
Blending
Board
Body
Book
Boom
Booming
Bootcamps
Boundaries
Boundless
Brain
Brainwave Sensing Gadgets
Brand
Breakthrough
Bridge

Brynjolfsson
Budding
Building
Business
Businesses
Calculations
Canvas
Capabilities
Capacity
Carbon Dioxide Filter
Career
Careers
Carnegie Learning's Mika
Carpentry
Carve
Catalyst
Catalyzing
Caution
Cave Art
Celebrities
Celebrity
Ceos
Chalkboards
Challenges
Change
Channels
Chapter
Charisma
Chatbots
Chatgpt 4
Chengdu
Child
Childhood
Children
China
Choice
Christianity
Chunk
Classroom
Clear
Clinical
Co-Creating
Code

Coding
Cognitive
Collaboration
Collaboration Tools
Collaborations
Combine
Comic Book Heroes
Comments
Commission
Commit
Communication
Communities
Community
Companies
Compassion
Compensation
Competencies
Competition
Complement
Complete
Completing
Complex
Complex Communication
Complex Equations
Complexity
Computational Conscience
Computers
Concepts
Conform
Conforming
Connect
Conservation
Consistency
Content
Content Creation
Content Diversity
Content Performance
Context
Contextual Reasoning
Contributions
Conventional
Conventions
Council
Countries
Coursera

Courses
COVID-19
Craftsmen
Creation
Creative
Creativity
Creators
Credibility
Credits
Criteria
Critical
Critical Thinking
Crops
Crucial
Cruise
Cultural Works
Culture
Curate
Curiosity
Currencies
Curricula
Curriculum
Customer Service
DALLE
Dance Routines
Daniel Goleman
Data
Data Processing
Daunting
Day
Decades
Decision-Making
Decisions
Dedication
Deep
Deep Blue
Deeper
Defer
Definitions
Delivery
Delta Airlines
Demand
Democratizes
Dermatology
Description

Design
Designer
Destiny
Determination
Determine
Detweiler
Develop
Developing
Devices
Digital
Digital Age
Digital Battlefield
Digital Content
Digital Domain
Digital Expanse
Digital Fluency
Digital Frontier
Digital Hunger
Digital Legacy
Digital Literacy
Digital Playgrounds
Digital Realm
Digital Tools
Digital World
Digitalization
Direct
Disciplines
Discussion
Diseases
Disruption
Disrupts
Diverse
Diverse
Division
Doctor
Domain
Domains
Dominance
Donations
Doors
Dr. Bertalan Mesko
Drake
Dream
Drive

Driven
Driving
Drones
Dyslexia
Early
Early 2000s
Easyjet
ECG
Economic Shifts
Economy
Education
Educational Landscape
Educators
Effectiveness
Effects
Efficiency
Efforts
Ekiti
Elderly
Electricity
Element
Elevate
Email
Embrace
Emergent AI
Emerging
Emotional Intelligence
Emotionally
Empathy
Empathy
Employees
Employment
Empower
Empowerment
Enable
Enables
Endeavors
Endorsements
Engage
Engagement
Engineers
Enlightenment
Enlightenment Era
Enrichment
Entertainment

Entrepreneurial Ventures
Entrepreneurs
Envy
Equipment
Equity
Era
Erik
Essence
Essential
Essentialism
Establishes
Establishing
Eternity
Ethical
Ethical Dilemma
Ethical Responsibility
Ethicists
Ethics
Evaluate
Evaluating
Evaluation
Evolution
Evolve
Evolving
Executive
Existence
Expand
Expanse
Experience
Experiences
Exploration
Expression
Extensive
Extravagant
Facilitate
Factories
Factories
Fairness
Faith
Fame
Fans
Fashion
Fashion Designer
Fields
Financial

Fine Art
Fire
Fiscal
Flexibility
Flexible
Focus
Focused
Followers
Following
Food
Force
Ford
Forgiveness
Form
Formal
Formative
Formula
Forum
Forums
Forward
Foxconn
Framework
Frameworks
Friction
From
Frontal Lobes
Fund
Fundamental
Fundamentals
Funding
Furnish
Future
Future Careers
Gadgets
Gain
Gaining
Gates
Gene Editing
General
General Electric
Generations
Generative
Generative AI
Generosity

Genesis
Geography
Germany
Gift
Giver Qualities
Giving Mentality
Glenda
Glimpse
Globe
Gold Standard
Goldmine
Google
Google Analytics
Google Labs' Magenta
Governance
Graphic Design
Gratitude
Ground
Grounding
Growth
Growth Mindset
Guidance
Hammer
Hands-On
Hard Work
Harness
Harris
Hashtags
Have
Healers
Healthcare
Heart
Helm
Henry
Highlights
Hiring Process
History
Holistic
Hollywood
Home Economics
Homes
How
Human
Human Agents

Human Bonds
Human Connection
Human Conscience
Human Food
Human Ingenuity
Human Intelligence
Human Relationships
Human Touch
Humanity
Human-Made Structures
Hybrid
IBM
IBM PC
IBM's Watson
Identity
Illustrates
Imagination
Imaginative
Immense
Impact
Importance
Impressions
Improve
Incentives
Including
Inclusion
Income
Incremental
Individual
Industry
Inequality
Influence
Influencer Marketing Industry
Influencers
Information
Ingenuity
Initiatives
Innovate
Innovation
Innovations
Innovators
Inquiry
Insights
Inspiration

Inspirational
Instagram
Institutions
Instruction
Instructional
Integration
Integrity
Intelligence
Intelligence Quotient
Interaction
Interactive
Interconnected
Interconnectedness
Interest
Interests
Internal Combustion Engine
Internet
Interpersonal Skills
Into
Intricacies
Introduction
Intuition
Inventions
Invest
Investments
Iphone
IQ
Izotope's Neutron And Ozone
James
James Clear
Jameson
Jeff
Jigsaw Puzzle
Job
Job Market
John Locke
John Lundgren
Journey
Joy
Judgment
Kasparov
Keith Richards
Ken Robinson
Key
Knowledge

Kritik
Kuka
Kurzgesagt
Labor
Labor Laws
Labor Unions
LANDR
Landscape
Laptops
Large-Frame
Lawmakers
Laws Of Physics
Leaders
Leadership
Leading
Lean Manufacturing
Learn
Learner
Learners
Learning
Lens
Lessons
Levy
Life Skills
Lifelong
Lifelong Learning
Life's Meaning
Like
Like-Minded
Likes
Limit
Limitations
Limited
Lines
Linguistic
Literacies
Literacy
Livelihoods
Lives
Livestreaming
Logical-Mathematical
Long-Term
Love
Loyalty
Luddite Movement

Luminaries
Luxury
Mac
Machine Learning
Machine Matcher
Machinery
Machines
Machines
Macintosh
Magenta
Magic
Magnates
Maintain
Major
Make
Makers
Making
Management
Manual Labor
Manufacturing
Marc
Markets
Martin
Martin Ford
Massive
Masterpieces
Mathematics
Mattan
Mcafee
Mckinsey
Meaningful Relationships
Media
Media Industry
Media Landscape
Medical AI
Medicine
Meet
Meeting
Memorization
Mentoring
Mentorship
Merchandise
Merits
Methods
Metrics

Michael Jackson
Microsoft
Microsoft Excel
Middle School
Midjourney
Milestone
Milestones
Mind
Minds
Mirror
Misconceptions
MITS
Mobility
Molly
Monetization
Monetization Maze
Moocs
Moral
Moral Wisdom
More
Most
Motivated
Motivations
Motor Skills
Move
MS-DOS
Multidimensional
Multilateral
Music
Must
Myth
Narrative
Narratives
Natural Language Processing
Natural Language Processing (NLP)
Nature
MacNaughton
Navigate
Need
Needs
Netscape Navigator
Neutron
Neville Goddard
New
Niche

Nigeria
Nigeria Bank
Noncognitive
Not
Novel
Now
Nuances
Numeracy
Nurture
Nurturing
Offer
Offers
Often
On
One
Online
Online Audience
Online Communities
Online Domain
Online Identity
Online Presence
Online World
Only
Open
Open Enrollment Program
Openness
Opportunities
Opportunity
Optimization
Order
Originality
Out
Outcomes
Outsourcing
Over
Oversight
Oxford Dictionary
Oxford University
Ozone
Pace
Paradigms
Partnerships
Passion
Passion-Fueled
Passions

Passive
Past
Path
Paths
Pathways
Pattern
Patterns
Paypal
Pearson
Peer
People
Perceptions
Perform
Performance
Personal Computers
Personalities
Personalize
Personalized
Perspective
Perspectives
Philanthropic
Philippines
Philosophy
Photographs
Photoshop AI
Physics
Physics
Pilots
Pivotal
Platforms
Play
Playgrounds
Please
Policies
Policy
Policymakers
Portfolio
Portfolios
Posts
Potential
Potential
Power
Powers
Practical Skills
Practices

Premium
Prepare
Presence
Pressure
Price
Primary
Principles
Principles
Priorities
Private
Proactive
Proactively
Problem-Solving
Productivity
Profession
Professional Opportunities
Professional World
Professional Writer
Professionals
Professors
Profit
Profound Spiritual Need
Programming
Programming
Programs
Progress
Projects
Promise
Prospective Employees
Provide
Prowess
Public
Purpose
Quality
Quantum Bits
Quantum Computing
Qubits
Questions
Quizlet
Raccoons
Race
Radiology
Rania
Rapidly
Rather

Raw Talent
Realm
Realms
Real-World
Recall
Receive
Recommend
Recommended
Reels
Reform
Reforming
Relationships
Remediation
Remove
Repetitive Tasks
Requires
Research
Research
Resemble AI
Reside
Resistance
Resourcefulness
Resources
Retail
Retraining
Revenue
Revolution
Revolutionize
Revolutionizing
Reward Mechanism
Rewards
Right
Rigid
Rise
Risk-Taking
Robotics
Robots
Roles
Rolling Stones
Room
Routines
Rules
Running Water
Rural Areas
Sacred Interdependence

Sacrifice
Safety
Safety Net
Sailors
Salesforce
Scaffolding
Scenario
Scenarios
School
Schooling
Science
Sciences
Screens
Sea
Second Industrial Revolution
Sector
Sectors
Seek
Seeking
Seem
Self-Awareness
Self-Directed
Self-Driving Cars
Self-Education
Self-Employment
Selfies
Self-Perceived
Send
Sensely
Sensory Information
Sent
Serve
Service Sector
SESAR JU
Set
Shadows
Shape
Shaping
Shared
Shelf-Life
Shelves
Shifts
Should
Showcases
Sidney

Siemens
Significance
Siloed
Similarly
Simon Sinek
Singapore
Siri
Skills
Skyline
Small
Smartphones
Snippets
Social
Social Hubs
Social Media
Social Media Education
Social Media Landscape
Social Media Platforms
Social Platforms
Social Space
Social Workers
Societal
Societal Level
Society
Socrates
Software
Solid
Solutions
Some
Souls
Space Exploration
Space-Time
Spatial Reasoning
Specialized
Spiderman
Spiritual Capacity
Spiritual Journey
Sponsors
Spotify
Spotlight
Stage
Stakeholders
Standard
Standardized

Standards

Stars
Start
Startups
Stay
Stories
Strategic
Strategies
Strategy
Streams
Stretch
Strong
Structures
Students
Study
Subjects
Success
Success On Creator Platforms
Successful
Support
Surtax
Swiss Army Knife
Sydney
Systems
Tablets
Tactical
Take
Taking
Talents
Tales
Tapestry
Task
Tasks
Tax
Teachers
Team
Teams
Technical
Techniques
Technological Change
Technological Innovation
Technologies
Technologists
Technology
Teenagers
Temporal Processing

Tesla
Testament
Text
Textiles
Thailand
The Weeknd
Themselves
Then
Theo
Theoretical
There
They
Think
Thinking
Thomas
Thought Experiment
Thoughtfully
Threshold
Thrive
Thriving
Through
Tide
Tiktok
Time
Time-Saving
Tips
Together
Tom Cruise
Toolbox
Tools
Trades
Tradition
Traditional
Traditional Employment
Traditional Paradigms
Traditional Pursuits
Trailblazers
Training
Transform
Transformations
Transformative Shift
Transforming
Transforms
Transition
Transparency

Travel
Trends
Trolley Problem
Trust
Truth
Tuition
Tuition-Free
Tutorials
Tweets
Two
Uber
Ubiquity
Udemy
Ultimately
Underserved
Unique
Uniqueness
University
Unstructured
Up
Upskilling
Urbanization
Us
Use
Value
Values
Velocity
Venture
Ventures
Versatile
Via
View
Viewpoints
Virtual Gifts
Virtual Realm
Virtual World
Visual Arts
Vital
Vocational
Voice Cloning
Vulnerability
Walls
Watch Hours
Waves
Wealth

Weaving
Web Browsers
Where
While
Wilderness
Window
Windows
Wisdom
With
Work Sectors
Workers
Workforce
Workshops
World
World Wide Web
Worldwide
Worth
Write
Writing
Writing Assistants
X
Youtube
Youtube Partner Program
Zeal

www.ingramcontent.com/pod-product-compliance
Lightning Source LLC
Chambersburg PA
CBHW062321290526
45794CB00005B/1853